Engineering Fluid Mechanics

Hongqing Song

Engineering Fluid Mechanics

冶 金 工 业 出 版 社
Metallurgical Industry Press

Hongqing Song
School of Civil and Resource Engineering
University of Science and Technology
  Beijing
Beijing
China

ISBN 978-981-13-4349-0   ISBN 978-981-13-0173-5   (eBook)
https://doi.org/10.1007/978-981-13-0173-5

Jointly published with Metallurgical Industry Press, Beijing, China

The print edition is not for sale in China Mainland. Customers from China Mainland please order the
print book from: Metallurgical Industry Press.

Printed on acid-free paper

This Springer imprint is published by the registered company Springer Nature Singapore Pte Ltd.
part of Springer Nature
The registered company address is: 152 Beach Road, #21-01/04 Gateway East, Singapore 189721,
Singapore

Preface

Fluid mechanics is a discipline which focuses on the fluid equilibrium and its movement. Based on the continuum model, we present the research of fluid mechanics to reader in this book. Contents of the book have been organized into broad topic areas: Introductory concepts and fluid statics (Chaps. 1, 2); Fluid dynamics, flow regimes, head loss, and pipe flow calculation (Chaps. 3, 4, 5); Fundamentals of fluid mechanics through porous media and corresponding applications (Chap. 6); Analysis of fluid machinery (Chap. 7); Similitude and dimensional analysis (Chap. 8); and Evaluation and analysis of CO_2 storage effect according to principles of seepage mechanics (Chap. 9).

This book provides many practical tables containing various parameters that are frequently used in engineering field. For example, Chap. 1 gives some tables about specific weights of various fluids at standard atmospheric pressure, along with water viscosity, water elastic modulus, and water expansibility coefficient at different conditions. In Chap. 2, the geometric properties of various structures are used for the determination of action location of resultant force, such as area, centroid coordinate, inertial moment, and so on. In Chap. 4, minor loss coefficients of different kinds of tubes are listed to facilitate utilization. Roughness coefficients of different kinds of surfaces, hydraulic conductivities of different kinds of soils, and ratio scales of similitude models are listed in Chaps. 5, 6, and 8, respectively.

This book also introduces two different ways to derive Darcy's law. The first method is based on the well-known Darcy's experiment, which is widely chosen in most textbooks. The other way presented in this book is to derive Darcy's law from N-S equation based on the connection between free flow and porous flow, which is helpful for readers to further study nanoscale flow. Furthermore, an advanced interdisciplinary case study on carbon dioxide capture and storage concerning energy and environment is presented. With this case study, we intend to provide a survey of the most increasing amounts of the world's energy needs from renewable resource.

At last, we select a number of novel and significant problem sets at the end of each chapter to help instructor assign homework and improve readers' ability to solve engineering problems.

This book is written for senior undergraduates, graduate students, lecturers, and researchers engaged in fluid mechanics and its application. It mainly introduces the basic theoretical knowledge with concise and understandable words, which is much easier for readers to understand. Meanwhile, this book describes clearly the difference and connection between free flow and porous flow, and incorporates the newest knowledge about nanoscale flow. What is more, the knowledge of carbon dioxide capture and storage is also introduced to bring readers' attention to environmental problems. The main benefit for reader is a sound introduction into all aspects of fluid mechanics covering all relevant subfields.

Beijing, China   Hongqing Song
March 2018

Contents

Introduction

1

Abstract

Fluid mechanics is a discipline which focuses on the fluid equilibrium and its movement. It has great significance on our daily lives. In this chapter, we will first introduce the developing trends and research methods of fluid mechanics. Then we will give definitions of continuum model for fluids. Finally, we will discuss major properties of fluid, such as specific weight, viscosity, compressibility and so on.

Keywords

Specific weight · Viscosity · Compressibility · Expansibility · Surface tension

1.1 Background

1.1.1 Definition of Fluid Mechanics

Fluid mechanics is a discipline which studies the fluid equilibrium and its movement law. In contrast to a solid, a fluid is a substance the particles of which easily moves and changes their relative position. More specifically, a fluid is defined as a substance that will deform continuously, that is, flows under the action of a sheer stress, no matter how small that shear stress may be. A fluid can either be a gas or a liquid.

The significance of fluid mechanics becomes apparent when we consider the vital role it plays in our everyday lives. For example, when we turn on our water taps, we activate flow in a complex hydraulic network of pipes, valves, and pumps. And our very lives depend on a very important fluid mechanic process—the flow of blood through our veins and arteries. Fluid mechanics is involved in some of the most significant environmental problems, such as air pollution and underground hazardous waste.

1.1.2   Trends in Fluid Mechanics

The science of fluid mechanics began with the need to control water for irrigation purposes in ancient Egypt, Mesopotamia, and India. Although these civilizations understood the nature of channel flow, there is no evidence that any quantitative relationships had been developed to guide them in their work. It was not until 250 B.C. that Archimedes discovered and recorded the principles of hydrostatics and flotation. Although the empirical understanding of hydrodynamics continued to improve with the development of fluid machinery, better sailing vessels, and more intricate canal systems, the fundamental principles of classical hydrodynamics were not set forth until the seventeenth and eighteenth centuries. Newton, Daniel Bernoulli, and Euler made the greatest contributions to these principles establishment.

Modern developments in fluid mechanics, as in all fields, involve the utilization of high-speed computers in the solution of problems. The ever-increasing speed and memory capacity of modern computers are leading to even more exciting applications of computers in fluid mechanics. Armed with more detailed measurements and numerical models, fluid mechanicians have developed higher levels of understanding that have led to sophisticated designs and applications of fluid systems.

1.1.3   Research Methods of Fluid Mechanics

There are three methods for fluid mechanics research: theoretical analysis, experimental test and numerical simulation. For theoretical analysis, there are three steps to investigate fluid flow. At first, several main factors affecting the flow problem are determined after analysis. Then the theoretical mathematic model is established with appropriate assumptions. Eventually the general solution for fluid movement is obtained by mathematical method utilization. For experimental test, the actual flow problem is summarized as a similar experimental model. With the combination of experimental equipment such as wind tunnel and water tunnel, the actual result can be deduced based on phenomena and data according to experiments. For numerical simulation, the calculation results based on theoretical analysis and experimental observation can be obtained with computing approaches, such as finite difference method and finite element method, etc. The numerical simulation is becoming popular with computing technology development since twenty-first century.

These three methods have their own advantages and disadvantages, and the best way is to complement each other to promote the development of fluid mechanics. With the development of modern measurement and computing technology, fluid mechanics will get largely promotion and facilitate for industrial development.

1.2 Continuum Model of Fluid

1.2.1 Fluid Particles

Similar to solid, fluid also has three basic properties: first, it consists of a large number of molecules; second, the molecules keep random thermal motion; third, there are molecular forces between molecules. From the view of microscopic level, a fluid molecule must have a certain shape, which means a fluid is not continuous, although the gap between molecules is very tiny.

However, fluid mechanics mainly focus on macroscopic mechanical movement of fluids rather than microscopic molecules, which is the statistical mean behavior of a large quantity of molecules. Therefore, the concepts of fluid particle and continuum must be adopted for fluid mechanic investigations.

Fluid particle is regarded as an aggregation of fluid molecules with small enough volume and definite macroscopic parameters value such as density, viscosity and velocity, etc.

1.2.2 Continuum Model of Fluid

Assuming that a fluid is composed of fluid particles rather than molecules, then the fluid is a continuum, which means that differential calculus is valid for all the analysis of fluid mechanics. Therefore, all physical properties of fluid particle, such as density, velocity, pressure, and temperature, are the continuous functions of space and time (x, y, z, t) [1–3]. The mathematical tools of continuous function and field theory can be utilized to solve flowing problems.

1.3 Main Properties of Fluid

1.3.1 Density and Specific Weight

Density is the mass of the substance per unit volume. It indicates the intensity of fluid in space, and usually it is denoted by the Greek letter ρ with unit kg/m^3 in SI.

For homogenous fluid, each point has the same density

$$\rho = \frac{m}{V} \tag{1.1}$$

For heterogeneous fluid, take an infinitesimal volume ΔV surrounding a certain point in space, the mass of fluid in it is Δm, then the ratio $\Delta V/\Delta m$ is the average density in volume ΔV. Let $\Delta V \to 0$, the limit of this ratio will be the density at the point inside ΔV.

$$\rho = \lim_{\Delta V \to 0} \frac{\Delta m}{\Delta V} = \frac{\mathrm{d}m}{\mathrm{d}V} \qquad (1.2)$$

The specific weight of a fluid is the gravity per unit volume, denoted by γ. For homogenous fluid

$$\gamma = \frac{G}{V} = \frac{mg}{V} = \rho g \qquad (1.3)$$

For heterogeneous fluid:

$$\gamma = \lim_{\Delta V \to 0} \frac{\Delta G}{\Delta V} = \frac{\mathrm{d}G}{\mathrm{d}V} \qquad (1.4)$$

The unit of specific weight in SI is N/m^3.

Different fluids have different density and specific weight. The density and specific weight of same fluid also vary with the temperature and pressure. Table 1.1 gives the density and specific weight of various fluids at standard atmospheric pressure. Table 1.2 gives the density and specific weight of water at standard atmospheric pressure with different temperature.

1.3.2 Viscosity

Viscosity is another important property of a fluid. Viscosity is a measure of the fluid's resistance to flow due to its internal friction. Viscosity is measured in two ways: dynamic (absolute) viscosity μ and kinematic viscosity υ. One must note that a fluid exhibits viscosity only when there is relative motion between fluid elements, or the fluid is in motion. A fluid at rest will not exhibit viscosity. The effect of viscosity indicates that it will restrict relative slip inside a fluid, thereby obstructing a fluid's flow. But this obstruction can only slow down the process of relative slip rather than eliminating this phenomenon, and this is an important feature of viscosity.

1.3.2.1   Newton's Viscosity Law

As shown in Fig. 1.1, two parallel plates are placed horizontally with a distance h. A certain liquid is filled between the two plates, then supposing that the lower plate is fixed and the upper moves with a uniform velocity v_0 to the right under the application of force F. According to the nonslip condition, a viscous fluid will stick to the solid boundary which the fluid attaches, so the fluid layer attaching to the upper plate moves at the velocity v_0 along the direction of x axis, and the fluid layer attaching to the lower plate stays at rest. Fluid between the two plates flows in the direction parallel to the plates, its velocity changes uniformly from zero of the lower

Table 1.1 Physical properties of various fluids at standard atmospheric pressure

Fluid	Temp. (°C)	Density (kg/m^3)	Specific weight (N/m^3)	Dynamic viscosity (Pa s)	Kinematic viscosity (m^2/s)
Distilled water	4	1000	9800	1.52×10^{-3}	1.52×10^{-6}
Seawater	20	1025	10,045	1.08×10^{-3}	1.05×10^{-6}
Carbon tetrachloride	20	1588	15,562	0.97×10^{-3}	0.61×10^{-6}
Gasoline	20	678	6644	0.29×10^{-3}	0.43×10^{-6}
Petroleum	20	856	8389	7.2×10^{-3}	8.4×10^{-6}
Lubricant	20	918	8996	440×10^{-3}	479×10^{-6}
Kerosene	20	808	7918	1.92×10^{-3}	2.4×10^{-6}
Alcohol	20	789	7732	1.19×10^{-3}	1.5×10^{-6}
Glycerol	20	1258	12,328	1490×10^{-3}	1184×10^{-6}
Turpentine	20	862	8448	1.49×10^{-3}	1.73×10^{-6}
Castor oil	20	960	9408	0.961×10^{-3}	1.00×10^{-6}
Benzene	20	895	8771	0.65×10^{-3}	0.73×10^{-6}
Mercury	0	13,600	133,280	1.70×10^{-3}	0.125×10^{-6}
Liquid hydrogen	−257	72	705.6	0.021×10^{-3}	0.29×10^{-6}
Liquid oxygen	−195	1206	11,819	82×10^{-3}	68×10^{-6}
Air	20	1.20	11.76	1.83×10^{-5}	1.53×10^{-5}
Oxygen	20	1.33	13.03	2.0×10^{-5}	1.5×10^{-5}
Hydrogen	20	0.0839	0.8222	0.9×10^{-5}	10.7×10^{-5}
Nitrogen	20	1.16	11.37	1.76×10^{-5}	1.52×10^{-5}
Carbon monoxide	20	1.16	11.37	1.82×10^{-5}	1.57×10^{-5}
Carbon dioxide	20	1.84	18.03	1.48×10^{-5}	0.8×10^{-5}
Helium	20	0.166	1.627	1.97×10^{-5}	11.8×10^{-5}
Methane	20	0.668	6.546	1.34×10^{-5}	2.0×10^{-5}

plate to v_0 of the upper plate. Thus, there is relative motion between every two fluid layers, and inner friction T will be generated on the interface. Assuming that the contacting area of the plate with the fluid is A. Shear stress in a fluid is defined as the inner friction per unit area, and is denoted with the symbol τ, so $\tau = T/A$.

Assuming that flowing velocity distribution complies with the relationship of linearism, as shown in Fig. 1.1. Experiments demonstrate that the magnitude of shear stress τ in a fluid is directly proportional to the velocity of upper plate, but is inversely proportional to the distance between the two plates, so we have

$$\tau = \mu \frac{v_0}{h} \qquad (1.5)$$

Table 1.2 Physical properties of water at different temperature (1 standard atmospheric pressure)

Temp. (°C)	Density (kg/m³)	Specific weight (N/m³)	Dynamic viscosity ×10⁻³ Pa s	Kinematic viscosity ×10⁻⁶ m²/s	Bulk modulus ×10⁹ N/m²	Surface tension (N/m)
0	999.9	9805	1.792	1.792	2.04	0.0762
5	1000.0	9806	1.519	1.519	2.06	0.0754
10	999.7	9803	1.308	1.308	2.11	0.0748
15	999.1	9798	1.140	1.141	2.14	0.0741
20	998.2	9789	1.005	1.007	2.20	0.0731
25	997.1	9779	0.894	0.897	2.22	0.0726
30	995.7	9767	0.801	0.804	2.23	0.0718
35	994.1	9752	0.723	0.727	2.24	0.0710
40	992.2	9737	0.656	0.661	2.27	0.0701
45	990.2	9720	0.599	0.650	2.29	0.0692
50	988.1	9697	0.549	0.556	2.30	0.0682
55	985.7	9679	0.506	0.513	2.31	0.0674
60	983.2	9658	0.469	0.477	2.28	0.0668
70	977.8	9600	0.406	0.415	2.25	0.0650
80	971.8	9557	0.357	0.367	2.21	0.0630
90	965.3	9499	0.317	0.328	2.16	0.0612
100	958.4	9438	0.284	0.296	2.07	0.0594

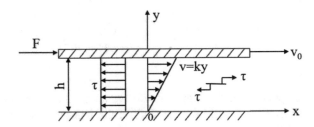

Fig. 1.1 Newton's viscosity law [4]

μ is a proportionality constant relevant to fluid property, and is called dynamic viscosity, and its unit is Pa s.

If velocity distribution does not comply with the relationship of linearism, as shown in Fig. 1.2, we have

$$\tau = \pm\mu\frac{du}{dy},$$ (1.6)

where $\frac{du}{dy}$ is called velocity gradient (shear rate).

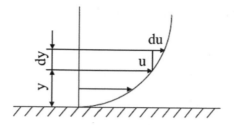

Fig. 1.2 Velocity gradient [5]

The above equation is the expression of Newton's viscosity law, which means that the shear stress is proportional to the velocity gradient of the fluid and is relevant to the property of the fluid.

Fluids for which shear stress and velocity gradient satisfy Newton's viscosity law are called Newtonian fluids. A lot of substances in nature, such as water, air, gases, and ordinary oils belong to Newtonian fluids. For some fluids, however, the shear stress may not be directly proportional to shear rate, and these fluids are classified as non-Newtonian fluids. As shown in Fig. 1.3, there are mainly three kinds of non-Newtonian fluids [6].

(1) Bingham plastic. Fluids that have a linear shear stress/shear strain relationship require a finite yield stress before they begin to flow. Several examples are clay suspensions, drilling mud, toothpaste.
(2) Shear thinning fluid. When the shear rate increases, the viscosity of a shear thinning fluid appears to decrease. The polymer solution is a common example.
(3) Shear thickening fluid. The viscosity of a shear thickening fluid appears to increase when the shear rate increases. Corn starch dissolved in water is a common example: when stirred slowly it looks milky, when stirred vigorously it feels like a very viscous liquid.

Fig. 1.3 Newtonian and non-Newtonian fluids

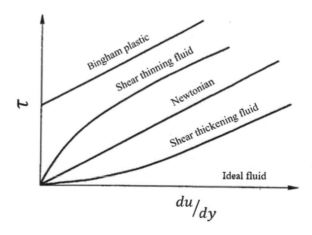

The kinematic viscosity v can be obtained by dividing the dynamic viscosity μ by density, that is

$$v = \frac{\mu}{\rho} \tag{1.7}$$

The unit of kinematic viscosity v is m^2/s.

1.3.2.2   Measurement of Viscosity

The measurement of viscosity is generally made with a device known as a viscometer. Various types of viscometers are available, such as U-tube viscometers, falling sphere viscometers and so on. For U-tube viscometers, as shown in Fig. 1.4, they are also known as glass capillary viscometers or Ostwald viscometers, named after Wilhelm Ostwald. Another version is the Ubbelohde viscometer, which consists of a U-shaped glass tube held vertically in a controlled temperature bath. In one arm of the U is a vertical section of precise narrow bore (the capillary). Above there is a bulb, with it is another bulb lower down on the other arm. In use, liquid is drawn into the upper bulb by suction, then allowed to flow down through the capillary into the lower bulb. Two marks (one above and one below the upper bulb) indicate a known volume. The time taken for the level of the liquid to pass between these marks is proportional to the kinematic viscosity.

Fig. 1.4   U-tube viscometer

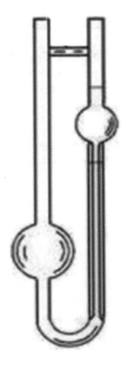

Most commercial units are provided with a conversion factor, or can be calibrated by a fluid of known properties. The time required for the test liquid to flow through a capillary of a known diameter of a certain factor between two marked points is measured. By multiplying the time taken by the factor of the viscometer, the kinematic viscosity is obtained.

Example 1.1 As shown in Fig. 1.5, the mass of a 1 cm height and $40 \times 45\,\mathrm{cm}^2$ bottom area wood board is 5 kg. It moves at a fixed velocity along a slope with lubricant. The wood velocity is $u = 1\,\mathrm{m/s}$, and the oil thickness is $d = 1\,\mathrm{mm}$. The oil velocity gradient caused by wood is linear. What is the dynamic viscosity μ of oil?
Solution

$$\theta = \arctan(5/12) = 22.62°$$

Since velocity is fixed, $a_s = 0$. Based on Newton's second law

$$\sum F_s = ma_s = 0$$

$$mg \sin \theta - \tau \cdot A = 0$$

Because velocity gradient is linear,

$$\tau = \mu du/dy = \mu u/\delta$$

Thus

$$\mu = mg \sin \theta \cdot \delta/Au = 0.105\,\mathrm{Pa\,s}$$

1.3.2.3 Variation Rules of Viscosity
The viscosity of fluid varies with temperature and pressure. Compared with the temperature, the effect of pressure is much smaller and is usually neglected.

Fig. 1.5 Example 1.1

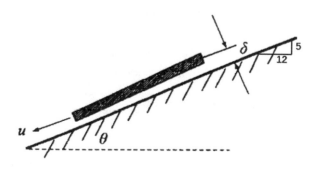

A number of empirical formulas reflecting viscosity variation with temperature are obtained from numerous experimental data.

Empirical formula about the relationship between liquid dynamic viscosity and temperature is as follows:

$$\mu = \mu_0 e^{-\lambda(t-t_0)}, \tag{1.8}$$

where μ_0 is liquid dynamic viscosity at t_0,Pa s; λ is a coefficient, which reflects the change rate of viscosity with different temperature.

Empirical formula about the relationship between gas dynamic viscosity and temperature is as follows:

$$\mu = \mu_0 \frac{1+\frac{C}{273}}{1+\frac{C}{T}} \sqrt{\frac{T}{273}}, \tag{1.9}$$

where μ_0 is gas dynamic viscosity at 0 °C, Pa s; T is gas thermodynamic temperature, K; C is a constant. The constants of ordinary gases are shown in Table 1.3.

For various liquids and gases, the relationship between dynamic viscosity μ and temperature is shown in Fig. 1.6, and the relationship between kinematic viscosity υ and temperature is shown in Fig. 1.7. The viscosity of water and air at different temperature and standard atmospheric pressure are shown in Table 1.4.

As shown in Figs. 1.6 and 1.7, Table 1.4, the variation rules of viscosity for liquids and gases are different when temperature changes. For example, the viscosity of liquid decreases while the gas viscosity increases with increase of temperature.

When temperature increases, distance between liquid molecules increases, and the attraction between molecules becomes weaker, so the viscosity of liquids will decrease significantly. As for gases, irregular motion of molecules intensifies with temperature increases, and momentum exchange becomes more frequent, thereby viscosity of gases will increase.

1.3.2.4   Ideal Fluid

An ideal fluid is usually defined as a fluid in which there is no friction, namely $\mu = \upsilon = 0$. Although such a fluid does not exist in reality, many fluids are approximate frictionless flow at sufficient distances from solid boundaries, so their behaviors can be characterized by assuming they act as ideal fluids.

Table 1.3   Values of C for various gases

Gas	Air	Hydrogen	Oxygen	Nitrogen	Steam	Carbon dioxide	Carbon monoxide
C	122	83	110	102	961	260	100

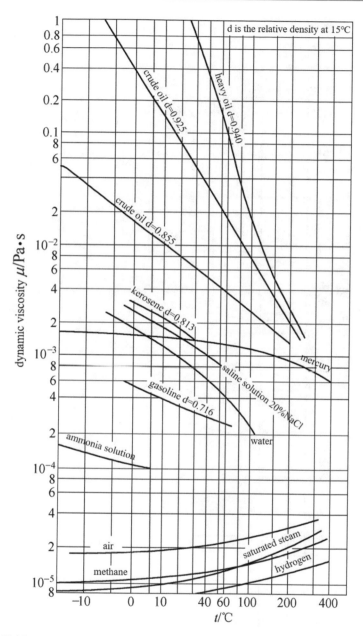

Fig. 1.6 Fluid dynamic viscosity [5]

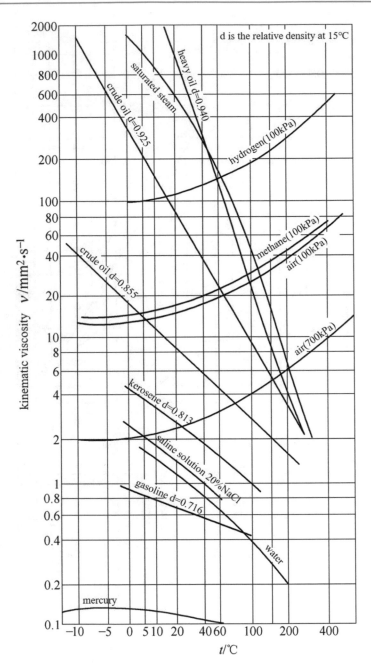

Fig. 1.7 Fluid kinematic viscosity [5]

Table 1.4 Viscosity of water and air at standard atmospheric pressure

Temp. t (°C)	Water		Air	
	μ (Pa s)	υ (m^2/s)	μ (Pa s)	υ (m^2/s)
0	1.792×10^{-3}	1.792×10^{-6}	0.0172×10^{-3}	13.7×10^{-6}
10	1.308×10^{-3}	1.308×10^{-6}	0.0178×10^{-3}	14.7×10^{-6}
20	1.005×10^{-3}	1.005×10^{-6}	0.0183×10^{-3}	15.3×10^{-6}
30	0.801×10^{-3}	0.801×10^{-6}	0.0187×10^{-3}	16.6×10^{-6}
40	0.656×10^{-3}	0.661×10^{-6}	0.0192×10^{-3}	17.6×10^{-6}
50	0.549×10^{-3}	0.556×10^{-6}	0.0196×10^{-3}	18.6×10^{-6}
60	0.469×10^{-3}	0.477×10^{-6}	0.0201×10^{-3}	19.6×10^{-6}
70	0.406×10^{-3}	0.415×10^{-6}	0.0204×10^{-3}	20.6×10^{-6}
80	0.357×10^{-3}	0.367×10^{-6}	0.0210×10^{-3}	21.7×10^{-6}
90	0.317×10^{-3}	0.328×10^{-6}	0.0216×10^{-3}	22.9×10^{-6}
100	0.284×10^{-3}	0.296×10^{-6}	0.0218×10^{-3}	23.6×10^{-6}

1.3.3   Compressibility and Expansibility

The density of a fluid is the function of temperature and pressure, thus the volume occupied by fluid will vary with pressure and temperature. The common sense is that volume reduces when pressure increases, while volume expands as temperature increases. This kind of attribute for a fluid is termed as compressibility and expansibility.

1.3.3.1   Compressibility

The compression coefficient C_L is defined as the relative change in volume caused by a unit change in pressure at constant temperature.

$$C_L = -\frac{\frac{\mathrm{d}v}{V}}{\mathrm{d}p} = -\frac{1}{V}\frac{\mathrm{d}V}{\mathrm{d}p}\,\mathrm{m}^2/\mathrm{N} \qquad (1.10)$$

or

$$C_L = \frac{1}{\rho}\frac{\mathrm{d}\rho}{\mathrm{d}p}\,\mathrm{m}^2/\mathrm{N}, \qquad (1.11)$$

where the minus is used to render the compression coefficient C_L positive, since relative volume change is always negative number.

The reciprocal of compression coefficient C_L is called the bulk modulus E.

$$E = \frac{1}{C_L}\,\mathrm{N}/\mathrm{m}^2 \qquad (1.12)$$

Table 1.5 Water's bulk modulus at different temperature and pressure

Temp. (°C)	Pressure/MPa				
	0.5	1	2	4	8
0	1.852×10^9	1.862×10^9	1.882×10^9	1.911×10^9	1.940×10^9
5	1.891×10^9	1.911×10^9	1.931×10^9	1.970×10^9	2.030×10^9
10	1.911×10^9	1.931×10^9	1.970×10^9	2.009×10^9	2.078×10^9
15	1.931×10^9	1.960×10^9	1.985×10^9	2.048×10^9	2.127×10^9
20	1.940×10^9	1.980×10^9	2.019×10^9	2.078×10^9	2.173×10^9

Table 1.5 shows the bulk modulus of water. It can be seen from Table 1.5 that the bulk modulus of water changes very little with pressure. Water is difficult to compress. In engineering applications, most liquids can be considered as incompressible fluids.

Example 1.2 Condensing a certain liquid in a container. When the pressure is 10^6 Pa, the volume of the liquid is 1L. If the pressure turns into 2×10^6 Pa, the volume of the liquid is 995 cm³. Find the liquid's bulk modulus.
Solution
According to Eq. (1.12)

$$E = \frac{1}{C_L} = -\frac{dp}{\frac{dv}{V}} = -\frac{2 \times 10^6 - 1 \times 10^6}{\frac{995-1000}{1000}} = 2 \times 10^8 \, \text{Pa}$$

1.3.3.2 Expansibility
The volume expansion coefficient β_t is defined as the relative change in volume caused by unit change in temperature at constant pressure.

$$\beta_t = \frac{\frac{dv}{V}}{dt} = \frac{1}{V}\frac{dV}{dt} \, 1/°C \qquad (1.13)$$

Table 1.6 shows the volume expansion coefficient of water. It can be seen from Table 1.6 that the expansibility of water is very small. Similar to the water, the other liquids also demonstrate small expansibility. Except for large changes in temperature, the expansibility of liquids can usually be ignored in engineering applications.

The volume of gases can change significantly with pressure and temperature. Near ambient temperature and at moderate pressure (<10 MPa), real gases can be

Table 1.6 Coefficient of volume expansion of water

Pressure/MPa	Temperature (°C)				
	1–10	10–20	40–50	60–70	90–100
0.1	0.14×10^{-4}	1.50×10^{-4}	4.22×10^{-4}	5.56×10^{-4}	7.19×10^{-4}
10	0.43×10^{-4}	1.65×10^{-4}	4.22×10^{-4}	5.48×10^{-4}	7.04×10^{-4}
20	0.72×10^{-4}	1.83×10^{-4}	4.26×10^{-4}	5.39×10^{-4}	–
50	1.49×10^{-4}	2.36×10^{-4}	4.29×10^{-4}	5.23×10^{-4}	6.61×10^{-4}
90	2.29×10^{-4}	2.89×10^{-4}	4.37×10^{-4}	5.14×10^{-4}	6.21×10^{-4}

considered as ideal gases. The following equation of state for ideal gases can be shown:

$$pV = mRT \text{ or } p = \rho RT, \tag{1.14}$$

where p is absolute pressure, Pa; T is absolute temperature (thermodynamic temperature), K; R is relevant to the relative molecular mass of the gas, and its magnitude may be expressed with the following equation:

$$R = \frac{\text{Molar gas constant}}{\text{Relative molecular mass}(M)} = \frac{8314}{M} \tag{1.15}$$

The unit of R is $N\,m/(kg\,K)$.

1.3.3.3 Compressible and Incompressible Fluids

Fluid mechanics deals with both incompressible and compressible fluids, that is, with liquids and gases of either constant or variable density. In reality, there is no incompressible fluid which is commonly used in case the density change with pressure is small enough to be neglected. This is usually the case with liquid.

In most cases, gas can be treated as compressible fluid, because its density always varies with the temperature and pressure.

1.3.4 Surface Tension

1.3.4.1 Surface Tension

Considering the following experiment, as shown in Fig. 1.8. A soap solution is used to form a soap film on a wire support. A cotton thread is draped across the soap film, as shown in Fig. 1.8a. If the soapy film ruptures on one side of the thread, the cotton thread will be pulled to one side. This is shown as Fig. 1.8b or c. This phenomenon shows evidence that there exists some forces (tension) within the soap film.

According to the theory of molecular attraction, molecules of liquid below the surface act on each other by forces that are equal in all directions. However, molecules near the surface have a greater attraction than those below the surface

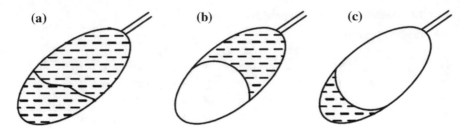

Fig. 1.8 Soap films

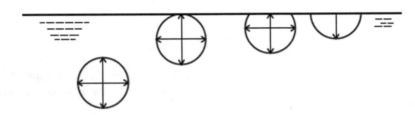

Fig. 1.9 Attraction between molecules near the surface

(see Fig. 1.9). This produces a surface on the liquid that acts like a stretched membrane. Because of this membrane effect, each portion of the liquid surface exerts "tension" on adjacent portions of the surface. These forces can act on any object in contact with the liquid surface.

This tension exists on the surface and its magnitude per unit length is defined as what is called surface tension, that is,

$$T = \sigma L, \tag{1.16}$$

where T is tension (force); L is length; σ is called surface tension, N/m. Table 1.7 gives values for surface tension of some liquids.

Table 1.7 Surface tension of liquids (20 °C, with air)

Liquids	Surface tension σ (N/m)	Liquids	Surface tension σ (N/m)
Alcohol	0.0223	Water	0.0731
Benzene	0.0289	Mercury	
Carbon tetrachloride	0.0267	With air	0.5137
Kerosene	0.0233–0.0321	With water	0.3926
Lubricant	0.0350–0.0379	With vacuum	0.4857
Crude oil	0.0233–0.0379		

1.3.4.2 Capillarity

The effect of surface tension can be studied in the case of capillary action in a small tube shown in Fig. 1.10a. When the end of a very small diameter tube is placed upright into a reservoir of water, a characteristic curved water surface forms between the interface of the water and air. The relatively great attraction of the water molecules for the glass causes the water surface to curve upward in the region of the glass wall. The surface tension forces within the fluid act around the circumference of the tube, in the direction indicated in Fig. 1.10.

From free body considerations, equating the lifting force created by surface tension to the gravity force leads to the balance as follows:

$$\pi d\sigma \cos\theta = \rho g h \pi d^2 / 4 \qquad (1.17)$$

Then, the capillary rise can be deduced from Eq. (1.17):

$$h = \frac{4\sigma \cos\theta}{\rho g d}, \qquad (1.18)$$

where σ is surface tension, namely the force per unit length; θ is the wetting angle between the fluid and the solid surface; ρ is the liquid density; d is the tube diameter; h is the height the liquid is lifted.

Equation (1.18) can be used to compute the capillary rise or capillary depression in a small diameter tube. Surface tension effects are generally negligible in most macroscopic engineering applications. However, these effects can be important in microscopic applications, in studies of nanotechnology, the formation of fluid droplets and bubbles.

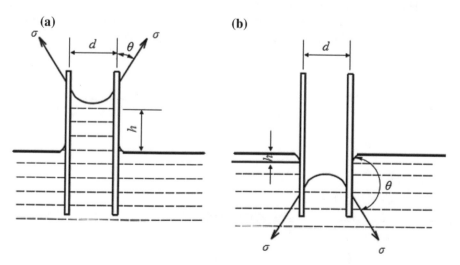

Fig. 1.10 Capillarity

1.4   Problems

1.1   Assuming that air's specific weight is $\gamma = 11.82\,\text{N/m}^3$, and dynamic viscosity is $\mu = 0.0183 \times 10^{-3}\,\text{Pa s}$. Find kinematic viscosity υ of air.

1.2   Find dynamic viscosity μ and kinematic viscosity υ of air when the temperature is 35 °C, and the pressure is 0.1 MPa.

1.3   Two parallel plates are placed horizontally with a distance $h = 10\,\text{mm}$. Castor oil (20 °C, $\mu = 0.972 \times 10^{-3}\,\text{Pa s}$) is filled between the two plates, then supposing that the lower plate is fixed and the upper moves with a uniform velocity ($v = 1.5\,\text{m/s}$). Find shear stress τ in the oil.

1.4   As shown in Fig. 1.11, a 1.5 m^2 bottom area wood board moves with a uniform velocity ($v = 16\,\text{m/s}$) on liquid surface which the thickness is 4 mm. Supposing that liquid velocity gradient caused by the board is linear, find,

   (1)   if the liquid is water at 20 °C, find the force F.
   (2)   if the liquid is oil at 20 °C ($\mu = 7.2 \times 10^{-3}\,\text{Pa s}$), find the force F.

1.5   As shown in Fig. 1.12, a horizontal interface 0-0 exists between two immiscible liquids. The dynamic viscosity of these two liquids are $\mu_1 = 0.14\,\text{Pa s}$ and $\mu_2 = 0.24\,\text{Pa s}$ respectively. The thickness is $\delta_1 = 0.8\,\text{mm}$ and $\delta_2 = 1.2\,\text{mm}$ respectively. Assuming that the velocity gradient is linear. Find the force F required to push a bottom area $A = 1000\,\text{cm}^2$ wood board to move with velocity $v_0 = 0.4\,\text{m/s}$ along the surface?

1.6   As shown in Fig. 1.13, a cone rotates around its vertical center axis with uniform velocity. Assuming that the gap is $\delta = 1\,\text{mm}$, and it is filled with lubricant which $\mu = 0.1\,\text{Pa s}$. In this figure, $R = 0.3\,\text{m}$, $H = 0.5\,\text{m}$, $\omega = 16\,\text{rad/s}$. Find the moment required to rotate the cone.

1.7   A tank is filled with oil, and the pressure is 49 kPa. The density of oil is 8900 kg/m^3. When the reduction of oil mass is 40 kg, the pressure correspondingly decreases to 9.8 kPa. Assuming that the bulk modulus of oil is $E = 1.32 \times 10^6\,\text{kN/m}^2$, find the volume of this oil tank.

Fig. 1.11 Problem 1.4

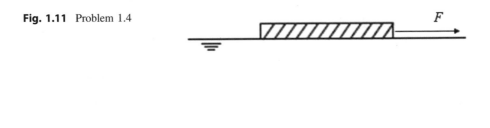

Fig. 1.12 Problem 1.5

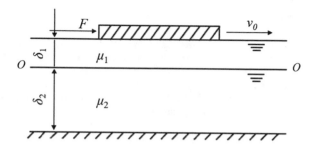

Fig. 1.13 Problem 1.6

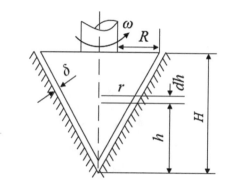

Fig. 1.14 Problem 1.9

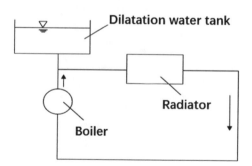

1.8 In a container of which the volume is $1.77\,\text{m}^3$, there are some CO with absolute pressure $P_0 = 103.4\,\text{kPa}$, and $T_0 = 21\,°\text{C}$. After the container is pumped into another $1.36\,\text{kg}$ CO, the temperature turns into $24\,°\text{C}$. Find the corresponding absolute pressure.

1.9 As shown in Fig. 1.14, in a heating system there is a dilatation water tank. The volume of the water in the system is $8\,\text{m}^3$. The largest temperature rise is $50\,°\text{C}$ and the volume expansion coefficient of water is $\alpha_v = 0.0005(1/°\text{C})$. Find the smallest volume of dilatation water tank.

References

1. White, F.M.: Fluid mechanics, 7th edn. McGraw-Hill, New York (2011)
2. Fox, R.W., McDonald, A.T., Pritchard, P.J.: Introduction to fluid mechanics, 8th edn, vol. 5. Wiley, Hoboken. New York (2011)
3. Panton, R.L.: Incompressible flow, 3rd edn. Wiley, New York (2005)
4. Xie, Z.: Engineering fluid mechanics, 4th edn. Metallurgical Industry Press, Beijing (2014)
5. Song, H.: Engineering fluid mechanics and environmental application. Metallurgical Industry Press, Beijing (2016)
6. Eric, F.R.: Rheology-theory and practice, vol. 3. Academic Press, New York (1960)

Fluid Statics

2

Abstract

Fluid statics focuses on equilibrium problems of forces exerting on a motionless fluid and corresponding application in practical situations. In this chapter, we will first discuss Eulerian equilibrium equation of fluids and its integral. In addition, we will introduce pressure calculation and measurement. Finally, according to the principles of pressure distribution, we can calculate forces acting on a plate.

Keywords

Surface force · Eulerian equilibrium · Pressure distribution · Absolute pressure
Vacuum degree · Pressure measurement

2.1 Forces on a Fluid

2.1.1 Mass Force

Mass force is a noncontact force acting on all fluid particles leading to a certain force field [1, 2]. The magnitude of mass force is proportional to the fluid mass or fluid volume. It is also called body force.

Unit mass force is defined as the mass force acting on unit mass, of which the unit is same with acceleration, m/s^2. If the fluid particle volume approaches zero, namely, $\Delta V \rightarrow 0$, then the mass force can be written as

$$\mathrm{d}F_m = \mathrm{d}m \cdot a_m = \mathrm{d}m(X\mathbf{i} + Y\mathbf{j} + Z\mathbf{k}), \tag{2.1}$$

© Metallurgical Industry Press, Beijing and Springer Nature Singapore Pte Ltd. 2018
H. Song, *Engineering Fluid Mechanics*,
https://doi.org/10.1007/978-981-13-0173-5_2

where dm is unit mass, a_m is acceleration, X, Y, Z are the components of acceleration or unit mass force on x, y, z axes respectively. General mass force includes gravity and inertial force.

2.1.2 Surface Force

Surface force is the force acting on the contact surface of fluid particles by the adjacent fluid layer or other fluid [3, 4]. Its value is proportional to contact area. According to the acting directions, surface forces can be divided into two categories: one is normal force which is perpendicular to acting surface such as pressure and normal stress, the other is shear force like shear stress which is parallel to acting surface.

2.2 Equilibrium Equation and Its Integral

2.2.1 Eulerian Equilibrium Equation

As we might all agree, the static fluid is in equilibrium. The total force involving the mass force and surface force in fluid element should equal zero, which can be used to establish the equilibrium equation and obtain the static pressure distribution.

As shown in Fig. 2.1, select a fluid cubic unit $abdcc\prime d\prime b\prime a\prime$ in a static fluid. Its length, width and height are dx, dy, dz, respectively. The pressure of the unit centroid $M(x, y, z)$ is $p(x, y, z)$. The forces acting on the fluid unit include

(1) Surface force. Since the fluid pressure is a continuous function of the position. Therefore, the pressure acting on surface ad and $a\prime d\prime$ in the x direction can be expanded as Taylor's series. Neglecting high order terms, the pressures of surface ad and $a\prime d\prime$ are $p + \frac{1}{2}\frac{\partial p}{\partial x}\,\mathrm{d}x$ and $p - \frac{1}{2}\frac{\partial p}{\partial x}\,\mathrm{d}x$, respectively. Similarly, the pressures of other surfaces are: $ac\prime$: $p - \frac{1}{2}\frac{\partial p}{\partial y}\,\mathrm{d}y$ and $bd\prime$: $p + \frac{1}{2}\frac{\partial p}{\partial y}\,\mathrm{d}y$ in the y direction; $a\prime b$: $p + \frac{1}{2}\frac{\partial p}{\partial z}\,\mathrm{d}z$ and $c\prime d$: $p - \frac{1}{2}\frac{\partial p}{\partial z}\,\mathrm{d}z$ in the z direction.

(2) Mass force. The components of the mass force on x, y, z axes are F_x, F_y and F_z, respectively:

$$F_x = \rho\,\mathrm{d}x\mathrm{d}y\mathrm{d}z\,X$$

$$F_y = \rho\,\mathrm{d}x\mathrm{d}y\mathrm{d}z\,Y$$

$$F_z = \rho\,\mathrm{d}x\mathrm{d}y\mathrm{d}z\,Z$$

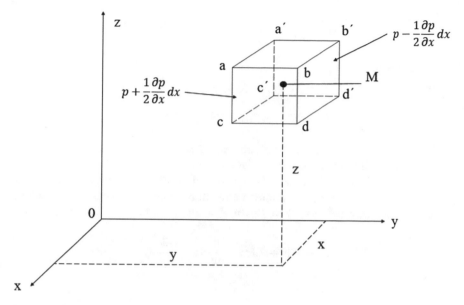

Fig. 2.1 Fluid cubic unit [5]

Since the fluid unit is in equilibrium, the total force in the x direction must be zero, thus

$$P_x + F_x = 0$$

Namely $-(p + \frac{1}{2}\frac{\partial p}{\partial x}dx)dydz + (p - \frac{1}{2}\frac{\partial p}{\partial x}dx)dydz + \rho dxdydzX = 0$

After simplification $-\frac{\partial p}{\partial x}dxdydz + \rho Xdxdydz = 0$

Similarly, we have

$$\left. \begin{array}{l} X - \frac{1}{\rho}\frac{\partial p}{\partial x} = 0 \\ Y - \frac{1}{\rho}\frac{\partial p}{\partial y} = 0 \\ Z - \frac{1}{\rho}\frac{\partial p}{\partial z} = 0 \end{array} \right\} \tag{2.2}$$

Equation (2.2) is the basic differential equation for fluids in equilibrium, which is first deduced by Euler in 1755, so it is also called Eulerian equilibrium equation. The above equation is suitable for any static fluid no matter how different force types and fluid properties are.

2.2.2 Integral of Equilibrium Equation

Multiplying Eq. (2.2) by dx, dy and dz respectively, then summing up the three equations and we have

$$\frac{\partial p}{\partial x}dx + \frac{\partial p}{\partial y}dy + \frac{\partial p}{\partial z}dz = \rho(Xdx + Ydy + zdz)$$

Since $p = p(x, y, z)$
then $dp = \frac{\partial p}{\partial x}dx + \frac{\partial p}{\partial y}dy + \frac{\partial p}{\partial z}dz$

$$dp = \rho(Xdx + Ydy + Zdz) \tag{2.3}$$

The above equation is the general differential equation for a fluid in equilibrium, which is also called pressure differential equation.

The left-hand side of Eq. (2.3) is the total derivative of pressure. Due to the uniqueness of integral result, the three terms on the right-hand side must also be a total derivative of a function $W = F(x, y, z)$, namely

$$dW = Xdx + Ydy + Zdz = \frac{\partial W}{\partial x}dx + \frac{\partial W}{\partial y}dy + \frac{\partial W}{\partial z}dz$$

Thus

$$X = \frac{\partial W}{\partial x}, Y = \frac{\partial W}{\partial y}, Z = \frac{\partial W}{\partial z} \tag{2.4}$$

Then Eq. (2.3) can be written as

$$dp = \rho dW \tag{2.5}$$

X, Y, Z in Eq. (2.4) are mass forces with potential which are equivalent to the gradient of a potential such as gravity, inertial force. Equation (2.5) is the total derivative equation of pressure p in static fluids, which shows the relationship between pressure and potential.

Integrating the both sides of Eq. (2.5), we have

$$p = \rho W + c,$$

where c is integral constant. If potential function W_0 and the pressure p_0 on free surface of the liquid in equilibrium are known, then $c = p_0 - \rho W_0$.

Therefore, the integral of Eulerian equilibrium equation can be expressed as

$$p = p_0 + \rho(W - W_0) \tag{2.6}$$

It can be known from Eq. (2.6) that if the potential function W is known, then the pressure of every point of a fluid in equilibrium can be determined. Equation (2.6) is significant to describe the pressure distribution in a static fluid.

2.2.3 Isobaric Surface

The curved or flat surface which consists of all points with the same pressure in a fluid is defined as isobaric surface. It is quite obvious that on an isobaric surface $dp = 0$, so Eq. (2.3) can be written as

$$X dx + Y dy + Z dz = 0 \qquad (2.7)$$

The isobaric surface has the following three properties:

(1) Isobaric surface is also the equal potential surface.

From Eq. (2.5), when $dp = 0$,

$$dW = 0, W = C$$

so the potential function is constant, namely, isobaric surface is also the equal potential surface.

(2) Mass force of each point on the isobaric surface is always perpendicular to isobaric surface.

Get a line segment $d\mathbf{l}$ arbitrarily on isobaric surface, it can be expressed as

$$d\mathbf{l} = dx\mathbf{i} + dy\mathbf{j} + dz\mathbf{k}$$

Equation (2.7) is the dot product of unit mass force and $d\mathbf{l}$. Since the dot product equals zero on an isobaric surface, mass force of each point on the isobaric surface is always perpendicular to isobaric surface.

(3) The interface between two kinds of immiscible liquid is isobaric surface.

As shown in Fig. 2.2, two kinds of immiscible liquids are in equilibrium in the vessel. Their densities are ρ_1 and ρ_2, respectively. If the interface $a - a$ between the two liquids is not an isobaric surface, then the pressure difference of point A and B can be obtained respectively from the two liquids:

$$\left. \begin{array}{l} dp = \rho_1 dW \\ dp = \rho_2 dW \end{array} \right\}$$

Since $\rho_1 \neq \rho_2$, the above two equations cannot be satisfied at the same time unless $dp = 0$ and $dW = 0$. Therefore, the interface $a - a$ must be an isobaric surface.

Fig. 2.2 Interface between
two static liquids

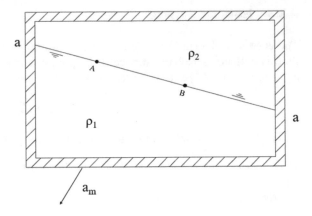

2.3   Basic Equation for Fluid Statics

2.3.1   Pressure Distribution in a Static Liquid

Figure 2.3 shows the coordinate system of a static liquid. The components of unit
mass force are $X = 0$, $Y = 0$ and $Z = -g$, and substituting them into Eq. (2.3), we
have

$$dp = \rho(-g dz) = -\gamma dz$$

Fig. 2.3   Static liquid

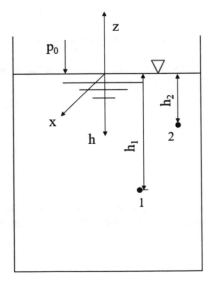

For homogeneous liquids ρ = constant, integrating the above formula, we have

$$p = -\gamma z + c \qquad (2.8)$$

$$z + \frac{p}{\gamma} = \text{constant} \qquad (2.9)$$

Equation (2.9) represents the pressure distribution in a static liquid, which is called the basic equation for fluid statics. It shows that the value of $z + \frac{p}{\gamma}$ at each point in the static liquid is identical. For example, for point 1 and 2 in Fig. 2.3, we have

$$z_1 + \frac{p_1}{\gamma} = z_2 + \frac{p_2}{\gamma}. \qquad (2.10)$$

2.3.2 Pressure Calculation in a Static Liquid

The constant c in Eq. (2.8) is the integral constant determined by boundary condition. If on the liquid surface $Z = 0$ and $p = p_0$, then it can be obtained from Eq. (2.8) that $c = p_0$.

$$\text{So} \quad p = p_0 - \gamma z \qquad (2.11)$$

If the direction of h is opposite to the z axis, then

$$p = p_0 + \gamma h \qquad (2.12)$$

The above equation is the formula of pressure at each point in a static liquid, which shows that the pressure at each point in a static liquid is the sum of the liquid surface pressure and the liquid gravity pressure γh. In homogeneous static liquid, the pressure at each position varies with its depth. The greater the depth h below the liquid surface is, the greater the pressure p is.

The isobaric surface of the fluid in equilibrium is perpendicular to mass force, and the mass force in a static liquid only shows gravity, so the isobaric surface of the static liquid must be horizontal.

For all kinds of communicating vessels containing homogeneous static liquid, if the depths of two points are the same, then the pressure of the two points must be identical.

Example 2.1 In the static liquid shown in Fig. 2.4, it is known that: p_a = 98 kPa, h_1 = 1 m, h_2 = 0.2 m, and the specific weight of oil and mercury are γ_{oil} = 7450 N/m^3 and γ_M = 133 kN/m^3, respectively. The height of point C and D are the same, try to determine the pressure at point C.

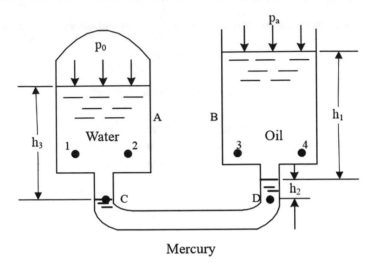

Fig. 2.4 Communicating vessel

Solution

From Eq. (2.12), the pressure of point D is:

$$p_D = p_a + \gamma_{\text{oil}}h_1 + \gamma_M h_2$$
$$= 98 + 7.45 \times 1 + 133 \times 0.2$$
$$= 132.05 \text{ kPa}$$

Since point C and D are of the same height in one continuous fluid, the pressure is identical, namely, $p_C = p_D = 132.05$ kPa.

2.3.3 Absolute Pressure, Relative Pressure, and Vacuum Degree

In engineering, there are two kinds of pressure representations. One is absolute pressure denoted as p which takes perfect vacuum as the measurement standard. The other is relative pressure denoted as p' which takes the atmospheric pressure p_a as the measurement standard, then $p' = p - p_a$. Generally, the pressure measured by a manometer is relative pressure which is also called gage pressure.

The relationship between absolute pressure and relative pressure is shown in Fig. 2.5. If $p > p_a$, then $p = p_a + p'$ and $p' = p - p_a$; if $p < p_a$, then $p = p_a - p_V$ and $p_V = p_a - p$, where p_V is called vacuum degree.

Example 2.2 Figure 2.6 is a sealed water tank. The distance between the water surface and plane $N - N$ is $h_1 = 0.2$ m, and the distance between $N - N$ and point M is $h_2 = 0.5$ m. Try to determine the absolute pressure and relative pressure of

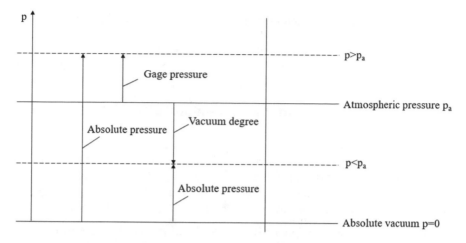

Fig. 2.5 Relationship among absolute pressure, gage pressure and vacuum degree [6]

Fig. 2.6 Sealed water tank

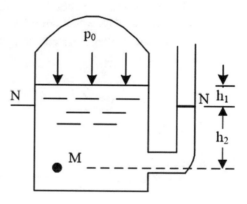

point M and find p_0. If there is vacuum on the water surface in the tank, try to determine the vacuum degree. The atmospheric pressure p_a is 101.3 kPa.
Solution

Since $N - N$ is isobaric surface, from Eq. (2.12) the pressure of point M can be obtained

$$p_M = p_a + \gamma h_2 = 101.3 + 9.8 \times 0.5$$
$$= 106.2 \text{ kPa}$$

$$p'_M = p_M - p_a = \gamma h_2 = 9.8 \times 0.5 = 4.9 \text{ kPa}$$

The absolute pressure on the water surface is

$$p_0 = p_M - \gamma(h_1 + h_2) = 106.2 - 9.8 \times (0.2 + 0.5) = 99.34 \text{ kPa}$$

Since $p_0 < p_a$, there is vacuum on the water surface. The vacuum degree is

$$p_v = p_a - p_0 = 101.3 - 99.34 = 1.96 \text{ kPa}.$$

2.3.4 Physical Meaning of Basic Equation for Fluid Statics

As shown in Fig. 2.7, taking plane $o - o$ as the datum reference, point A and B are connected with two piezometers opened to the atmosphere. The distance between point A and plane $o - o$ is z_A, and the distance between point B and plane $o - o$ is z_B. z_A and z_B are called elevation heads, namely, the potential energy per unit weight of the liquid relative to datum reference $o - o$. The water heights of the two piezometers are $\frac{p'_A}{\gamma}$ and $\frac{p'_B}{\gamma}$ respectively, which named relative pressure head.

Point C and D are connected with two closed piezometers. The distance between point C and plane $o - o$ is z_C, and the distance between point D and plane $o - o$ is z_D. z_C and z_D are also the elevation heads. The water heights of the two piezometers are $\frac{p_C}{\gamma}$ and $\frac{p_D}{\gamma}$ respectively, which named absolute pressure head. Pressure head can reflect the "pressure" energy per unit weight of fluid.

Figure 2.7 shows that

$$z_A + \frac{p'_A}{\gamma} = z_B + \frac{p'_B}{\gamma}, z_C + \frac{p_C}{\gamma} = z_D + \frac{p_D}{\gamma}$$

Fig. 2.7 Physical meaning of basic equations for fluid statics

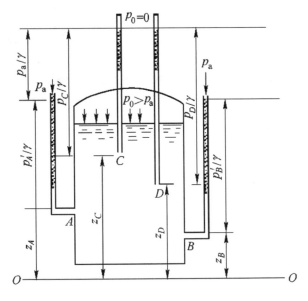

$$\text{or } z_A + \frac{p'_A}{\gamma} + \frac{p_a}{\gamma} = z_B + \frac{p'_B}{\gamma} + \frac{p_a}{\gamma} = z_C + \frac{p_C}{\gamma} = z_D + \frac{p_D}{\gamma}$$

Therefore, the total energy at each point in a static fluid is identical.

2.4   Pressure Measurement

2.4.1   Units of Pressure

The fundamental unit of pressure in SI system is Pascal (Pa), which is one Newton per square meter (N/m^2). Since Pa is too small, larger units kPa and MPa are used. Also, 1 kPa $= 10^3$ Pa and 1 MPa $= 10^6$ Pa.

Pressure is equivalent to a height of liquid column with constant density. Usually it is more convenient to express a pressure by the height of liquid column, rather than by the unit N/m^2. For example, mmH$_2$O or mmHg is just the common pressure unit.

Atmospheric pressure can also be used as a standard for pressure measurement

$$1 \text{ standard atmospheric pressure(atm)} = 760 \text{ mmHg} = 1.1013 \times 10^5 \text{ Pa}$$

To simplify engineering calculation, engineering atmospheric pressure is often used as a standard

$$1 \text{ Engineering atmospheric pressure(at)} = 9.8 \times 10^4 \text{ Pa} = 735.6 \text{ mmHg}$$
$$= 10 \text{ mH}_2\text{O}$$

Example 2.3 The height of water column caused by the pressure of a certain point is 6 m, what is the relative pressure of this point? How much standard atmospheric pressure and engineering atmospheric pressure does it equal respectively?

Solution

The relative pressure of this point is

$$p = \gamma h = 9800 \times 6 = 58800 \text{ N/m}^2 = 58.8 \text{ kN/m}^2$$

For standard atmospheric pressure

$$\frac{p}{p_{\text{atm}}} = \frac{58800}{1.013 \times 10^5} = 0.58$$

For engineering atmospheric pressure

$$\frac{p}{p_{\text{at}}} = \frac{58800}{98000} = 0.59 \,.$$

2.4.2 Pressure Measuring Instruments

(1) Piezometer

Basically, this method utilizes the change in pressure with height of a fluid column to evaluate pressure. As shown in Fig. 2.8, under the action of pressure at point A, the liquid in the piezometer is in equilibrium and its height is $h_A = \frac{p'_A}{\gamma}$, so the pressure at point A can be obtained.

(2) U-tube manometer

As shown in Fig. 2.9, U-tube manometer is a U-tube with a scale plate and can be conveniently utilized to measure pressure in a tank or pipeline [7].

Fig. 2.8 Piezometer

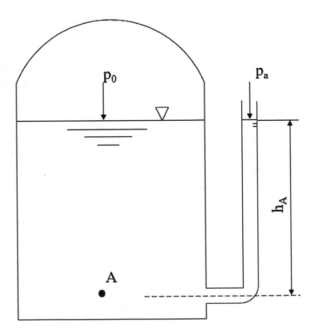

Fig. 2.9 U-tube manometer

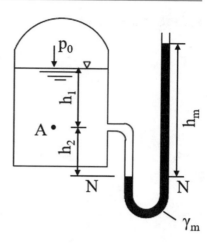

Since the interface is an isobaric surface, on the left side of the U-tube, we have

$$p_N = p_0 + \gamma(h_1 + h_2)$$

On the right side of U-tube, we have

$$p_N = p_a + \gamma_m h_m$$

Therefore

$$p_0 + \gamma(h_1 + h_2) = p_a + \gamma_m h_m$$

$$p_0 = p_a + \gamma_m h_m - \gamma(h_1 + h_2)$$

$$p_A = p_0 + \gamma h_1 = p_a + \gamma_m h_m - \gamma h_2$$

Then p_0 and p_A can be obtained after the measurement of h_1, h_2 and h_m.

(3) Cup manometer

The cup manometer is based on U-tube manometer, as shown in Fig. 2.10. It consists of a metal cup filled with mercury and a glass tube fixed on a scale plate. The specific weights of water and mercury are γ_W and γ_M, respectively. Then the absolute pressure at point C is

$$p_C = p_a + \gamma_M h - \gamma_W L \qquad (2.13)$$

Fig. 2.10 Cup manometer

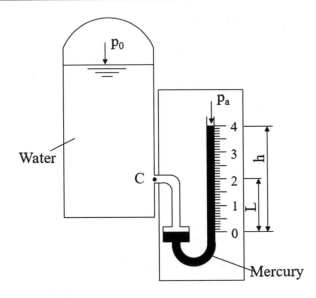

(4)   Differential manometer

In engineering, sometimes there is no need to know the pressure at a certain point but to know the pressure difference between two certain points, and the apparatus to measure pressure difference is called differential manometer. Figure 2.11 is a differential manometer for measuring the pressure difference between point A and B. The height difference of the mercury surface due to the pressure difference is Δh, so the pressure difference between point A and B is

Fig. 2.11 Differential manometer

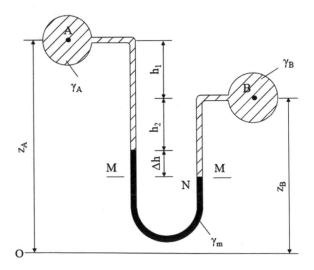

$$p_B - p_A = \gamma_A(h_1 + h_2) + \gamma_m \Delta h - \gamma_B(h_2 + \Delta h) \qquad (2.14)$$

If A and B are both water, then

$$p_B - p_A = \gamma_w h_1 + 12.6\, \gamma_w \Delta h = \gamma_w(z_A - z_B) + 12.6\, \gamma_w \Delta h \qquad (2.15)$$

Example 2.4 As shown in Fig. 2.9, a mercury U-tube manometer is attached to the container. It is known that $h_m = 1$ m, $h_1 = 0.3$ m and $h_2 = 0.4$ m, then what is the relative pressure of the liquid surface in the container? How much engineering atmospheric pressure does it equal?

Solution

The relative pressure on the liquid surface is

$$p_0 = \gamma_m h_m - \gamma(h_1 + h_2) = 133280 \times 1 - 9800 \times (0.3 + 0.4) = 126420 \text{ Pa}$$
$$= 126.4 \text{ kPa}$$

In terms of engineering atmospheric pressure

$$\frac{p_0}{p_{at}} = \frac{126420}{98000} = 1.29$$

Example 2.5 As shown in Fig. 2.12, a mercury U-tube manometer is attached to a gas–water separator. If $\Delta h = 200$ mm, then what is the height H of the water surface in the separator?

Solution

The vacuum degree on the water surface is

$$p_v = \gamma_M \Delta h = 133280 \times 0.2 = 26656 \text{ N/m}^2$$

Establish the basic equation for fluid statics between point A and B

$$0 + \frac{p_a}{\gamma} = H + \frac{p_B}{\gamma}$$

namely,

$$0 + \frac{p_a}{\gamma} = H + \frac{p_a - p_v}{\gamma}$$

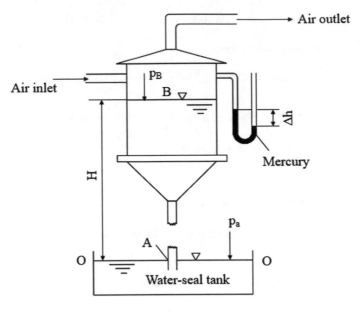

Fig. 2.12 Gas water separator

Therefore,

$$H = \frac{p_v}{\gamma} = \frac{26656}{9800} = 2.72 \text{ m.}$$

2.5   Forces Exerting on a Plate by Static Fluid

2.5.1   Resultant Force

Figure 2.13 shows a coordinate system and the angle α between an inclined plate and horizontal surface.

The resultant force exerting on the plate is the sum of the static pressure exerting on the plate, so the direction of the resultant force is perpendicular to the plate. Take an element dA from the plate, the distance between the liquid surface and its centroid is h, and the pressure on its centroid is

$$p = p_0 + \gamma h$$

So the resultant force exerting on the element is

$$dP = (p_0 + \gamma h)dA \tag{2.16}$$

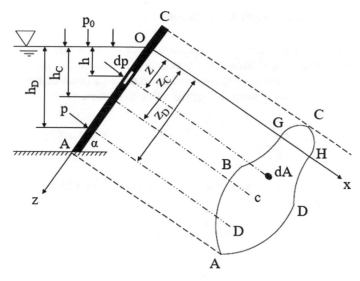

Fig. 2.13 Force exerting on plane area [5]

It can be known from Fig. 2.13 $h = z \sin \alpha$, so the resultant force exerting on the plate is

$$
P = \int_A (p_0 + \gamma h)\mathrm{d}A = \int_A (p_0 + \gamma z \sin \alpha)\mathrm{d}A
$$

$$
= p_0 A + \gamma \sin \alpha \int_A z \mathrm{d}A \tag{2.17}
$$

According to theoretical mechanics, $\int_A z\mathrm{d}A$ is the static moment of the area GBADH about x axis, and its value is $z_c A$, where z_c is the distance between the centroid c of the area A and x axis. Therefore

$$
P = p_0 A + \gamma \sin \alpha z_c A = p_0 A + \gamma h_c A, \tag{2.18}
$$

where h_c is the depth of the centroid of the area GBADH below the water surface.

For plate GBADH, its left side and right side are both under the action of p_0, so

$$
P = \gamma h_c A \tag{2.19}
$$

Regardless of the shape of the plate and its inclined angle, the magnitude of hydrostatic force exerting on a plate equals the product of plate centroid pressure and plate area on contact in terms of Eq. (2.19). The direction of the force is perpendicular to the plate.

2.5.2   The Action Location of Resultant Force

Assuming that the action location of resultant force shows as z_D, and the distance between the location of resultant force and liquid surface is h_D. According to theoretical mechanics, the moment of resultant force to an axis equals the sum of the moment of force components about the same axis, then

$$Pz_D = \int_A \gamma hz\,dA = \int_A \gamma z^2 \sin\alpha\,dA = \gamma \sin\alpha \int_A z^2\,dA, \qquad (2.20)$$

where $\int_A z^2 dA = I_x$ is the inertia moment of the area GBADH about x axis. The resultant force is $P = \gamma h_c A$, so

$$\gamma h_c A z_D = \gamma \sin\alpha I_x$$

$$z_D = \frac{\sin\alpha I_x}{h_c A} \qquad (2.21)$$

According to the parallel axis theorem for inertia moment: $I_x = I_c + z_c^2 A$. I_c is the inertia moment of the plate on contact about an axis which is parallel to x axis and through point c. Thus

$$z_D = \frac{\sin\alpha(I_c + z_c^2 A)}{h_c A} = \frac{\sin\alpha(I_c + z_c^2 A)}{z_c \sin\alpha A} = z_c + \frac{I_c}{z_c A} \qquad (2.22)$$

namely

$$z_D = z_c + \frac{I_c}{z_c A} \qquad (2.23)$$

The action location of resultant force is always below the centroid c according to above formula.

The geometric properties of several common planar graphs are shown in Table 2.1.

Example 2.6 Figure 2.14 shows a pool with a sluice gate. It's known $B = 2$ m and $h = 1.5$ m. Try to determine the magnitude and the action location of the resultant force exerting on the sluice gate.

Solution

Since $z_c = h_c = \frac{1}{2}h$ and $A = Bh$, from Eq. (2.19), we have

Table 2.1 The geometric properties of several common planar graphs [5]

Plane shape		Area A	Centroid coordinate z_c	Moment of inertia I_c
Rectangle		bh	$\frac{1}{2}h$	$\frac{1}{12}bh^3$
Triangle		$\frac{1}{2}bh$	$\frac{2}{3}h$	$\frac{1}{36}bh^3$
Circle		$\frac{1}{4}\pi d^2$	$\frac{d}{2}$	$\frac{\pi}{64}d^4$
Semicircle		$\frac{1}{8}\pi d^2$	$\frac{2d}{3\pi}$	$\frac{1}{16}\left(\frac{\pi}{8}-\frac{8}{9\pi}\right)$
Trapezoid		$\frac{h}{2}(a+b)$	$\frac{h}{3}\left(\frac{a+2b}{a+b}\right)$	$\frac{h^3}{36}\left(\frac{a^2+4ab+b^2}{a+b}\right)$
Ellipse		$\frac{\pi}{4}bh$	$\frac{h}{2}$	$\frac{\pi}{64}bh^3$

Fig. 2.14 Sluice gate

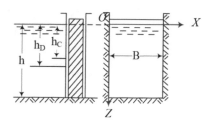

$$P = \gamma h_c A = \gamma \cdot \frac{1}{2}h \cdot Bh$$

$$= 9800 \times \frac{1}{2} \times 1.5 \times 2 \times 1.5$$

$$= 22050 \text{ N} = 22.05 \text{ kN}$$

From Table 2.1, we have

$$I_c = \frac{1}{12}Bh^3$$

From Eq. (2.23), the action location of the resultant force is

$$z_D = z_c + \frac{I_c}{z_cA} = h_c + \frac{I_c}{h_cA} = \frac{1}{2}h + \frac{\frac{1}{12}Bh^3}{\frac{1}{2}h \cdot Bh} = \frac{1}{2} \times 1.5 + \frac{\frac{1}{12} \times 2 \times 1.5^3}{\frac{1}{2} \times 1.5 \times 2 \times 1.5} = 1 \text{ m}$$

Example 2.7 Figure 2.15 shows an inclined sluice gate AB, and its width B is 1 m (perpendicular to the picture). It is known that $H = 3$ m and $h = 1$ m. Neglecting the gate weight and the friction, determine the vertical force needed to lift the gate.
Solution
From Eq. (2.19), the total pressure acting on the gate by the liquid is

$$P = \gamma h_cA = \gamma \cdot \frac{1}{2}H \cdot B \cdot \frac{H}{\sin 60°}$$
$$= 9800 \times \frac{1}{2} \times 3 \times 1 \times \frac{3}{\sin 60°}$$
$$= 50922 \text{ N} = 50.92 \text{ kN}$$

From Eq. (2.23), the distance between the action location of total pressure and point A is

$$l = \frac{h}{\sin 60°} + \left(z_c + \frac{I_c}{z_cA}\right)$$
$$= \frac{h}{\sin 60°} + \left(\frac{\frac{1}{2}H}{\sin 60°} + \frac{\frac{1}{12}B\left(\frac{H}{\sin 60°}\right)^3}{\frac{1}{2} \times \frac{H}{\sin 60°} \times B \times \frac{H}{\sin 60°}}\right)$$
$$= \frac{h}{\sin 60°} + \frac{H}{2\sin 60°} + \frac{H}{6\sin 60°} = 3.464 \text{ m}$$

Fig. 2.15 Inclined sluice gate

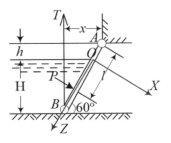

From Fig. 2.15, we have

$$x = \frac{H+h}{\tan 60°} = \frac{3+1}{\tan 60°} = 2.31 \text{ m}$$

According to the moment balance, we have

$$\sum M_A = Pl - Tx = 0$$

Therefore

$$T = \frac{Pl}{x} = \frac{50.92 \times 3.464}{2.31} = 76.36 \text{ kN}.$$

2.6 Problems

2.1 A diver works with a depth of 15 m underwater, try to determine the pressure exerting on the diver.

2.2 Figure 2.16 shows a closed vessel. It is known that $h = 1.5$ m, $p_a = 101.3$ kPa and $h' - 1.2$ m, try to determine the pressure p_0 on the water surface.

2.3. The pressure at a certain point in a water container is 1.5 mH$_2$O, try to determine the absolute pressure and relative pressure at this point. (Expressing them by mH$_2$O and mHg, respectively)

Fig. 2.16 Problem 2.2

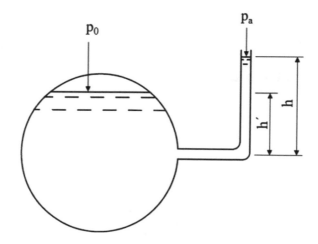

2.4. A container filled with gas is connected with a U-tube manometer containing
water, as shown in Fig. 2.17. If $h_v = 0.3$ m, try to determine the relative
pressure of the gas and its vacuum degree.

2.5. A water container is connected with a U-tube manometer containing mercury,
and relevant data are shown in Fig. 2.18. Try to determine the absolute
pressure and relative pressure at point M.

Fig. 2.17 Problem 2.4

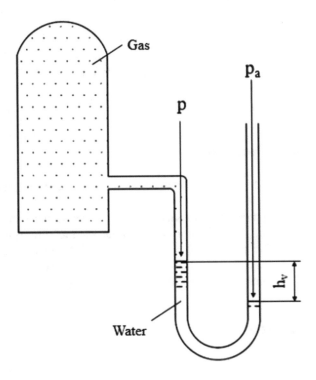

Fig. 2.18 Problem 2.5

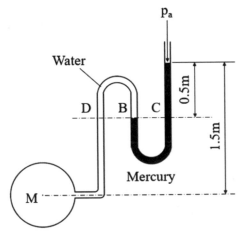

2.6. The air container is connected with two mercury U-tube manometers, and the specific weight of mercury is $\gamma_M = 133$ kN/m^3. The height difference of the open U-tube manometer is $h_1 = 30$ cm, as shown in Fig. 2.19. What is the height difference h_2 of the closed U-tube manometer? (the effect of gas specific weight is negligible.)

2.7. As shown in Fig. 2.20, container A is filled with water, and container B is filled with alcohol. The specific weight of alcohol is 8 kN/m^3. Using mercury U-tube manometer to measure the pressure difference of point A and B. It is known that $h_1 = 0.3$ m, $h = 0.3$ m and $h_2 = 0.25$ m, try to determine the pressure difference.

Fig. 2.19 Problem 2.6

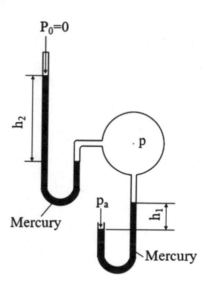

Fig. 2.20 Problem 2.7

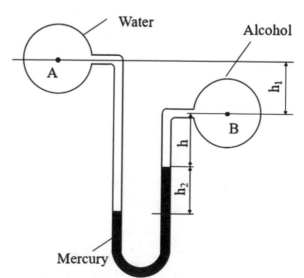

2.8. As shown in Fig. 2.21, the relative pressure A of coal gas in the pipe on the first floor is 100 mm H_2O. The height difference between the eighth and the first floor is $H = 32$ m. What is the relative pressure of coal gas in the pipe on the eighth floor? The densities of air and gas are assumed to be constants, and the coal gas specific weight is $\gamma_G = 4.9$ N/m^3.

2.9. The readings of mercury manometers are shown in Fig. 2.22, try to determine the relative pressure of point A in the water tank. (the heights are all relative to the ground, in unit of m)

2.10. Figure 2.23 shows a U-tube manometer used to measure the small pressure difference of the two water pipes. The top of the tube is filled with a liquid which is lighter than water and immiscible with water.

Fig. 2.21 Problem 2.8

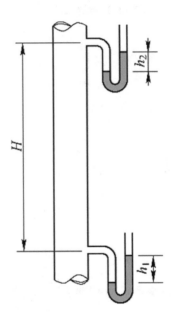

Fig. 2.22 Problem 2.9

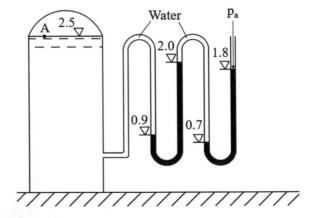

Fig. 2.23 Problem 2.10

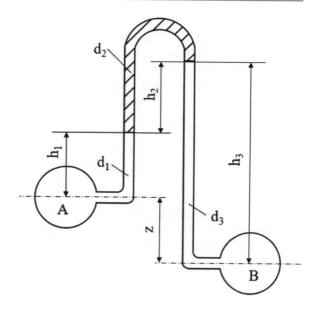

(1) It is known that the relative densities of the liquid in tube A and B are $d_1 = d_3 = 1$, and the relative density of the liquid in the U-tube manometer is $d_2 = 0.95$. If $h_1 = h_2 = 0.3$ m and $h_3 = 1$ m, try to determine the pressure difference $p_B - p_A$.

(2) If the pressure difference of the two pipes is $p_B - p_A = 3825.9$ Pa, try to determine the values of h_1, h_2, h_3 and z.

(3) Try to determine the pressure difference $p_B - p_A$ if $h_2 = 0$.

(4) If $d_2 = 0.6$ and $p_B - p_A = 0$, try to determine the values of h_1, h_2 and h_3.

2.11. As shown in Fig. 2.24, assuming that $p_2 = 98$ kN/m^2, $D = 100$ mm and $d = 30$ mm, try to determine the pressure p_1 needed to produce a force $F = 7840$ N in the horizontal direction. Assuming the total friction is 10% of the pressure F exerting on the piston.

Fig. 2.24 Problem 2.11

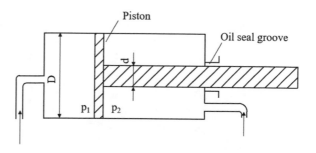

2.12. Figure 2.25 shows a cone-shaped water container. It is known that $D = 1\,\mathrm{m}$, $d = 0.5\,\mathrm{m}$ and $h = 2\,\mathrm{m}$. There is an object with $G = 3.2\,\mathrm{kN}$ on the lid. Determine the total pressure acting on the container bottom.

2.13. As shown in Fig. 2.26, there is a rectangular plate with inclined angle $60°$ underwater. The dimensions of this plate are: height 1.5 m, width 1.2 m. Try to determine the magnitude and action location h_D of the total pressure P acting on the plate.

2.14. As shown in Fig. 2.27, a circular sluice gate is placed at the bottom of a pool with an inclined angel $\theta = 60°$, and its diameter is $d = 1\,\mathrm{m}$. The water depth is $h = 2\,\mathrm{m}$. Determine the magnitude and action location h_D of the total pressure acting on the gate.

2.15. As shown in Fig. 2.28, the width of the navigation lock is $B = 25\,\mathrm{m}$, and the water level of the upstream and the downstream are $H_1 = 63\,\mathrm{m}$ and $H_2 = 48\,\mathrm{m}$ respectively. The lock consists of two rectangular gates; try to determine the magnitude and action location of the pressure acting on each gate.

Fig. 2.25 Problem 2.12

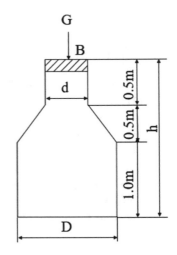

Fig. 2.26 Problem 2.13

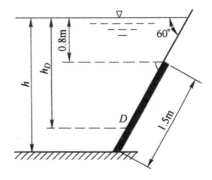

Fig. 2.27 Problem 2.14

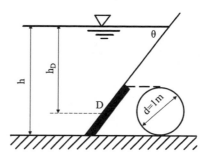

Fig. 2.28 Problem 2.15

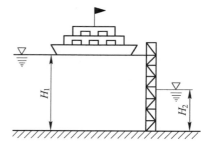

References

1. Roy, D.N.: Applied Fluid Mechanics. Ellis Horwood Limited (1988)
2. Shaughnessy, E.J., Katz, I.M., Schaffer, J.P.: Introduction to Fluid Mechanics. Oxford University Press Inc, Oxford (2005)
3. Potter, M.C., Wiggert, D.C., Hondzo, M.: Mechanics of Fluids, 2nd edn. Prentice-Hall Inc, Englewood Cliffs, NJ (1997)
4. Graebel, W.P.: Engineering Fluid Mechanics. Taylor & Francis Publishers, Abigdon (2001)
5. Song, H.: Engineering Fluid Mechanics and Environmental Application. Metallurgical Industry Press, Beijing (2016)
6. Xie, Z.: Engineering Fluid Mechanics, 4th edn. Metallurgical Industry Press, Beijing (2014)
7. Tabeling, P.: Introduction to Microfluidics. Oxford University Press Inc., New York (2005)

Fluid Dynamics

3

Abstract

Fluid dynamics, the study of fluid flow, is an important subset of fluid mechanics. It is based on the conservation laws of mass and momentum, moment of momentum to study the velocity and pressure of fluid flow, influential factors and their applications. In this chapter, we will first introduce the studying approaches (Lagrangian approach and Eulerian approach) and basic concepts of fluid flow. Then we will present continuity equation of fluid motion. Furthermore, we also show differential equations of motion for inviscid/viscous fluid and their corresponding Bernoulli's integral. Finally, we discuss momentum equation for steady flow and its application.

Keywords

Lagrangian approach · Steady flow · Streamline · Continuity equation
Bernoulli's integral · Pitot tube

3.1 Approaches Describing the Motion of Fluids

Generally, there are two approaches to describe the motion of a fluid for fluid mechanics investigation, which are Lagrangian approach and Eulerian approach.

3.1.1 Lagrangian Approach

The Lagrangian approach describes the motions and properties changes of an individual fluid particle with time. In order to distinguish one particle from another, the position (a, b, c) is conveniently utilized to describe a particular particle for all time.

© Metallurgical Industry Press, Beijing and Springer Nature Singapore Pte Ltd. 2018
H. Song, *Engineering Fluid Mechanics*,
https://doi.org/10.1007/978-981-13-0173-5_3

The position of a particle at each time can be expressed as

$$\left. \begin{array}{l} x = f_1(a,b,c,t) \\ y = f_2(a,b,c,t) \\ z = f_3(a,b,c,t) \end{array} \right\} \tag{3.1}$$

For a given position $(a,\ b,\ c)$, the above equations refer to the path for a particular fluid particle.

In this case, the particle velocity is obtained by differentiating the particle's position vector with respect to time. In the Cartesian coordinate system, the particle velocity is expressed as

$$\left. \begin{array}{l} u_x = \frac{\partial x}{\partial t} = \frac{\partial f_1(a,b,c,t)}{\partial t} \\ u_y = \frac{\partial y}{\partial t} = \frac{\partial f_2(a,b,c,t)}{\partial t} \\ u_z = \frac{\partial z}{\partial t} = \frac{\partial f_3(a,b,c,t)}{\partial t} \end{array} \right\} \tag{3.2}$$

And the acceleration will be

$$\left. \begin{array}{l} a_x = \frac{\partial^2 x}{\partial t^2} = \frac{\partial^2 f_1(a,b,c,t)}{\partial t^2} \\ a_y = \frac{\partial^2 y}{\partial t^2} = \frac{\partial^2 f_2(a,b,c,t)}{\partial t^2} \\ a_z = \frac{\partial^2 z}{\partial t^2} = \frac{\partial^2 f_3(a,b,c,t)}{\partial t^2} \end{array} \right\} \tag{3.3}$$

3.1.2   Eulerian Approach

A fluid is a continuum consisting of numerous fluid particles. The space filled with flowing fluids is called the flow field.

The Eulerian approach is to focus on a certain point in space, and consider the motion of fluid particles that pass through that point as time goes on. In a more general sense, the Eulerian approach describes the whole flow field by variation of fluid parameters with time at a certain point and nearby point in space, such as velocity and pressure.

In this case, the fluid parameters will depend on the point in space and time. Take velocity as an example. It can be expressed as follows:

$$\left. \begin{array}{l} u_x = F_1(x,y,z,t) \\ u_y = F_2(x,y,z,t) \\ u_z = F_3(x,y,z,t) \end{array} \right\} \tag{3.4}$$

And the acceleration will be

$$
\left.
\begin{aligned}
a_x &= \frac{\mathrm{d}u_x}{\mathrm{d}t} = \frac{\mathrm{d}F_1(x,y,z,t)}{\mathrm{d}t} \\
a_y &= \frac{\mathrm{d}u_y}{\mathrm{d}t} = \frac{\mathrm{d}F_2(x,y,z,t)}{\mathrm{d}t} \\
a_z &= \frac{\mathrm{d}u_z}{\mathrm{d}t} = \frac{\mathrm{d}F_3(x,y,z,t)}{\mathrm{d}t}
\end{aligned}
\right\}
\tag{3.5}
$$

The pressure and density can be expressed as $p = F_4(x, y, z, t)$ and $\rho = F_5(x, y, z, t)$, respectively.

With the variation of time, the fluid particle will move from one point to another, which means that the position of a fluid particle is also a function of time.

By using the chain rule for differentiation of a multivariable function, we can express the acceleration of a fluid particle in the x-direction as

$$
\begin{aligned}
a_x &= \frac{\mathrm{d}u_x}{\mathrm{d}t} = \frac{\partial u_x}{\partial t} + \frac{\partial u_x}{\partial x}\frac{\mathrm{d}x}{\mathrm{d}t} + \frac{\partial u_x}{\partial y}\frac{\mathrm{d}y}{\mathrm{d}t} + \frac{\partial u_x}{\partial z}\frac{\mathrm{d}z}{\mathrm{d}t} \\
&= \frac{\partial u_x}{\partial t} + u_x\frac{\partial u_x}{\partial x} + u_y\frac{\partial u_x}{\partial y} + u_z\frac{\partial u_x}{\partial z}
\end{aligned}
\tag{3.6}
$$

And similarly

$$
a_y = \frac{\mathrm{d}u_y}{\mathrm{d}t} = \frac{\partial u_y}{\partial t} + u_x\frac{\partial u_y}{\partial x} + u_y\frac{\partial u_y}{\partial y} + u_z\frac{\partial u_y}{\partial z}
\tag{3.7}
$$

$$
a_z = \frac{\mathrm{d}u_z}{\mathrm{d}t} = \frac{\partial u_z}{\partial t} + u_x\frac{\partial u_z}{\partial x} + u_y\frac{\partial u_z}{\partial y} + u_z\frac{\partial u_z}{\partial z}
\tag{3.8}
$$

or in the vectorial form

$$
\mathbf{a} = \frac{\mathrm{d}\mathbf{u}}{\mathrm{d}t} = \frac{\partial \mathbf{u}}{\mathrm{d}t} + (\mathbf{u} \cdot \nabla)\mathbf{u},
\tag{3.9}
$$

where $\nabla = \frac{\partial}{\partial x}\mathbf{i} + \frac{\partial}{\partial y}\mathbf{j} + \frac{\partial}{\partial z}\mathbf{k}$, known as the gradient, or Hamilton operator.

The terms on the right-hand side (RHS) of the Eq. (3.9) involve two different types: the change of velocity with respect to time and the change of velocity with respect to position.

Terms of the first type $\frac{\partial \mathbf{u}}{\mathrm{d}t}$ are called local accelerations or time-varying accelerations. Local acceleration terms occur only when a flow field is unsteady. In a steady flow, the local acceleration is zero. The last three terms $(\mathbf{u} \cdot \nabla)\mathbf{u}$ on the RHS of the Eq. (3.9) are called convective accelerations or space-varying accelerations. Convective acceleration occurs when velocity is a function of position in a flow field. In uniform flows, the convective acceleration is zero.

Example 3.1 Assuming that the velocity components of a fluid particle in a flow field are $u_x = kx, u_y = -ky, u_z = 0$. What is the acceleration?

Solution

The velocity is

$$u = \sqrt{u_x^2 + u_y^2} = k\sqrt{x^2 + y^2} = kr$$

The acceleration in the *x*-direction is given by

$$a_x = \frac{du_x}{dt} = u_x \frac{\partial u_x}{\partial x} = k^2 x$$

And similarly

$$a_y = \frac{du_y}{dt} = u_y \frac{\partial u_y}{\partial y} = k^2 y$$

$$a_z = 0$$

The acceleration is

$$a = \sqrt{a_x^2 + a_y^2} = k^2 \sqrt{x^2 + y^2} = k^2 r$$

3.2   Classification and Basic Concepts of Fluid Flow

3.2.1   Classification of Fluid Flow

There are many different types of flow and can be classified from different points of view, including

(1) Based on fluid characteristics or properties

Inviscid fluids are ideal fluids in which viscosity effect is ignored. The flow of fluids without viscosity is termed ideal flow, while the opposite is called viscous flow.

Flow can also be classified as an incompressible (liquid) and compressible (gas) flow. All fluids are compressible to some extent, that is, changes in pressure or temperature cause changes in density. However, in many situations changes in pressure and temperature are sufficiently small that the changes in density are negligible. In this case the flow can be modeled as an incompressible flow. Otherwise the more general compressible flow equations must be taken into consideration.

(2) Based upon the state of flow

According to the difference of flow state, fluid flow can be classified as: steady flow and unsteady flow, uniform flow and nonuniform flow, rotational flow and irrotational flow, laminar flow and turbulent flow, subsonic flow, transonic flow and supersonic flow, etc.

For example, a flow that is not a function of time is called steady flow. Time dependent flow is known as unsteady (also called transient) flow. Steady-state flow refers to the condition where the fluid properties at a point in the system do not change over time, as shown in Fig. 3.1. The fluid parameters for steady flow can be expressed as

$$\left.\begin{array}{l} u = f_1(x, y, z) \\ p = f_2(x, y, z) \\ \rho = f_3(x, y, z) \end{array}\right\} \tag{3.10}$$

(3) Based on the number of space variables

According to how many space variables flow parameters are dependent on fluid flow may be classified as one-dimensional, two-dimensional, and three-dimensional flow. This type of classification is applicable in all coordinate systems.

3.2.2 Basic Concepts of Fluid Flow

(1) Path line

Path lines are the trajectories that individual fluid particles follow. These can be thought of as "recording" the path of a fluid element in the flow over a certain

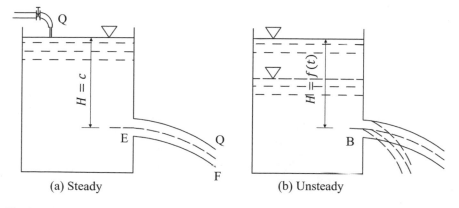

(a) Steady   (b) Unsteady

Fig. 3.1 Steady flow and unsteady flow

period. The direction the path takes will be determined by the streamlines of the fluid at each moment in time. It is a concept relevant to Lagrangian approach.

As shown in Fig. 3.2, after a short time Δt, the fluid particle M has moved from point A to point B, and the curve AB is the path line of particle M. Let dl denote the little displacement of particle M in time dt, then its velocity can be expressed as

$$u = \frac{dl}{dt}$$

Its components on x, y, and z axes are

$$\left.\begin{aligned} u_x &= \frac{dx}{dt} \\ u_y &= \frac{dy}{dt} \\ u_z &= \frac{dz}{dt} \end{aligned}\right\}, \tag{3.11}$$

where dx, dy and dz are the projection of dl in the coordinate system. It can be obtained from Eq. (3.11) that

$$\frac{dx}{u_x} = \frac{dy}{u_y} = \frac{dz}{u_z} = dt \tag{3.12}$$

Equation (3.12) is called the differential equation of path line.

(2) Streamline

Streamlines are a family of curves that are instantaneously tangent to the velocity vector of the flow as shown in Fig. 3.3[1]. A streamline exhibits the velocity direction of different massless fluid element at the same instant, and it is a concept relevant to Eulerian approach.

Fig. 3.2 Pathline

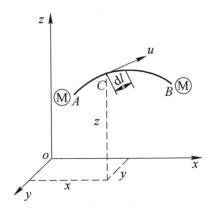

Fig. 3.3 Streamline

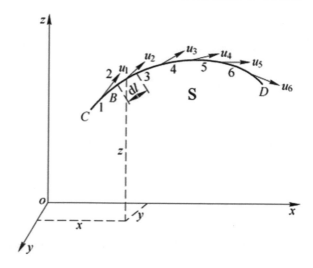

Path line and streamline coincide together in a steady flow, but usually do not in an unsteady flow. Since the velocity at any point in flow field is unique and determined, there is no possibility for two different streamlines to intersect at a point.

Streamline is a geometric description of velocity field. If the velocity distribution of a flow field is known, streamline equation can be derived from the differential equation of streamline. The derivation of the differential equation of streamline is shown as follows.

Assuming the curve s in Fig. 3.3 is a streamline, the velocity of a fluid particle at any point A on the streamline is **u**, and then take an elementary streamline ds at point A. By the definition of streamline, ds//**u** must be satisfied, namely,

$$\mathbf{u} \times d\mathbf{s} = 0 \tag{3.13}$$

Since the direction of ds and **u** is the same, their components on x, y, and z axes are in proportion correspondingly, therefore

$$\frac{dx}{u_x} = \frac{dy}{u_y} = \frac{dz}{u_z} \tag{3.14}$$

Equations (3.13) or (3.14) is called the differential equation of streamline.

Example 3.2 Assuming that the velocity components in a flow field are $u_x = x + t$, $u_y = -y + t$, $u_z = 0$, find the path line and streamline which pass through point $(-1, -1)$ when $t = 0$.

Solution

(1) The differential equation of path line in the problem is

$$\frac{dx}{dt} = x+t, \quad \frac{dy}{dt} = -y+t,$$

where time t should be regarded as an independent variable. Integrating the above equations, we have

$$x = c_1 e^t - t - 1, \quad y = c_2 e^{-t} + t - 1$$

When $t = 0$, substitute $x = y = -1$ into the above equations, get $c_1 = c_2 = 0$. Thus

$$x + y = -2$$

(2) The differential equation of streamline in the problem is

$$\frac{dx}{x+t} = \frac{dy}{-y+t},$$

where time t should be regarded as a constant. Integrating the above equations, we have

$$(x+t)(-y+t) = c$$

When $t = 0$, substitute $x = y = -1$ into the above equations, get $c = -1$. Thus

$$xy = 1$$

3.2.3   Cross Section, Velocity, and Flow Rate

The area which is perpendicular to all streamlines of the elementary stream beam (or stream tube control volume) is called cross section, as shown in Fig. 3.4.

According to the definition of cross section, the cross section is a flat surface if streamlines are parallel lines, otherwise it is a curved surface with various shapes.

Fig. 3.4 Cross section

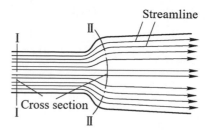

Fluid velocity differs with its position on the cross section. The velocity at a given point on cross section is usually denoted by symbol u, m/s. So the flow rate passing through a differential area dA is

$$dQ = udA \qquad (3.15)$$

The unit is m³/s (L/s) or mass flow rate kg/s.

And the total flow rate can be obtained by integrating the above equation over the entire area A of flow

$$Q = \int_Q dQ = \int_A udA \qquad (3.16)$$

The average velocity can be obtained by dividing the total flow rate by the area of cross section, namely

$$v = \frac{Q}{A} = \frac{\int_A udA}{A} \qquad (3.17)$$

3.3   Continuity Equation of Fluid Motion

The flow field is regarded as full of fluid particles without any forms of voids or fractures, and this property is called continuity [1–3]. The continuity equation is based upon the law of mass conservation as it applies to the flow of fluids. In this section, we first discuss the continuity equation in rectangular coordinate system, then apply it to elementary flow beam and total flow.

3.3.1   The Continuity Equation in Rectangular Coordinate System

Take arbitrarily an element with point M as the center, and its dimensions are $\delta x, \delta y, \delta z$ respectively, as shown in Fig. 3.5. The coordinates of point M are x, y, z,

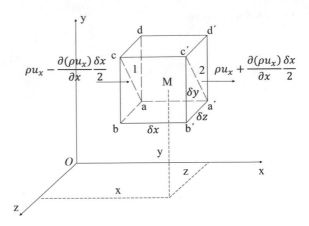

Fig. 3.5 An element in fluid motion

and the velocity and density of point M at time t are u and ρ, respectively. Since the element is infinitesimal, the velocity and density of every point at time t can be expanded as Taylor series neglecting high order terms. For example, the velocity of point 2 is $u_x + \frac{\partial u_x}{\partial x} \cdot \frac{\delta x}{2}$, and so on.

Now considering the fluid mass flowing through parallel surfaces $a\ b\ c\ d$ and $a'b'c'd'$ during time δt. Since the time and the element are infinitesimal, the velocities of all points are considered to be the same and distribute uniformly. Thus, the inflow mass through surface $a\ b\ c\ d$ is

$$\left[\rho u_x - \frac{\partial(\rho u_x)}{\partial x} \cdot \frac{\delta x}{2} \right] \delta y \delta z \delta t$$

The outflow mass through surface $a'b'c'd'$ is

$$\left[\rho u_x + \frac{\partial(\rho u_x)}{\partial x} \cdot \frac{\delta x}{2} \right] \delta y \delta z \delta t$$

Their difference, namely the net inflow mass is

$$-\frac{\partial(\rho u_x)}{\partial x} \cdot \delta x \delta y \delta z \delta t$$

Similarly, the net inflow mass in the y and z direction is respectively

$$-\frac{\partial(\rho u_y)}{\partial y} \cdot \delta y \delta x \delta z \delta t \quad \text{and} \quad -\frac{\partial(\rho u_z)}{\partial z} \cdot \delta z \delta x \delta y \delta t$$

The fluid mass increment in the element during time δt resulting from the change of fluid density in the element is

$$\left(\frac{\partial \rho}{\partial t}\delta t\right)\delta x\delta y\delta z$$

According to the law of mass conservation, the algebraic sum of three net inflow mass above must equal the fluid mass increment in the element, thus

$$-\left[\frac{\partial(\rho u_x)}{\partial x}+\frac{\partial(\rho u_y)}{\partial y}+\frac{\partial(\rho u_z)}{\partial z}\right]\cdot\delta x\delta y\delta z\delta t=\frac{\partial \rho}{\partial t}\delta t\delta x\delta y\delta z$$

Rewrite the above equation as

$$\frac{\partial \rho}{\partial t}+\frac{\partial(\rho u_x)}{\partial x}+\frac{\partial(\rho u_y)}{\partial y}+\frac{\partial(\rho u_z)}{\partial z}=0 \qquad (3.18)$$

This is the continuity equation for compressible three-dimensional flow. For compressible steady flow, the continuity equation is

$$\frac{\partial(\rho u_x)}{\partial x}+\frac{\partial(\rho u_y)}{\partial y}+\frac{\partial(\rho u_z)}{\partial z}=0 \qquad (3.19)$$

For incompressible fluid $\rho=$ constant, no matter the flow is steady or unsteady, Eq. (3.19) can be simplified as

$$\frac{\partial u_x}{\partial x}+\frac{\partial u_y}{\partial y}+\frac{\partial u_z}{\partial z}=0 \qquad (3.20)$$

Continuity equation establishes the relationship between velocity and density in a flow, which is a necessary condition for the existence of a given velocity field.

3.3.2 Continuity Equations for Elementary Flow Beam and Total Flow

3.3.2.1 Continuity Equation for Elementary Flow Beam

Considering an elementary flow beam shown in Fig. 3.6, the cross sections are dA_1 and dA_2. The corresponding velocities are u_1 and u_2, and the corresponding densities are ρ_1 and ρ_2. For compressible steady flow, the shape of flow beam will not change with time, which means that no fluid particles pass through the beam surface. During time dt, the inflow mass of dA_1 is $dM_1=\rho_1 u_1 dA_1 dt$, and the outflow mass of dA_2 is $dM_2=\rho_2 u_2 dA_2 dt$.

Fig. 3.6 Flow beam and
total flow [4]

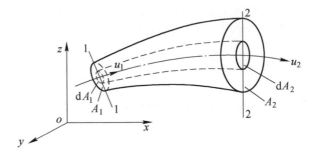

According to the law of mass conservation, the inflow must equal the outflow, thus

$$dM_1 = dM_2$$

Namely

$$\rho_1 u_1 dA_1 = \rho_2 u_2 dA_2 \tag{3.21}$$

For incompressible fluid $\rho_1 = \rho_2$, so

$$u_1 dA_1 = u_2 dA_2, \quad \text{namely} \quad dQ_1 = dQ_2 \tag{3.22}$$

This is the continuity equation of elementary flow beam for incompressible steady flow, which states that the flow rate of any cross section of the flow beam is the same.

3.3.2.2 Continuity Equation of Total Flow

Integrating Eq. (3.21) over corresponding cross section, we have

$$\int_{A_1} \rho_1 u_1 dA_1 = \int_{A_2} \rho_2 u_2 dA_2$$

Combining Eq. (3.17), the above equation can be rewritten as

$$\rho_{1m} v_1 A_1 = \rho_{2m} v_2 A_2$$

namely

$$\rho_{1m} Q_1 = \rho_{2m} Q_2 \tag{3.23}$$

where ρ_{1m} and ρ_{2m} are the average density of cross sections 1 and 2, respectively. Equation (3.23) is the continuity equation of total flow.

For incompressible fluids, it can be written as

$$Q_1 = Q_2 \quad \text{or} \quad A_1 v_1 = A_2 v_2 \tag{3.24}$$

Equation (3.24) indicates that the area of cross section is inversely proportional to the velocity in fluid motion which satisfies continuity. For example, the nozzle of the fire hose and the sedimentation pool in wastewater treatment are both applications of this rule in engineering.

The above continuity equation of total flow is derived from condition that the flow rate is constant along the flow channel. If there is another inflow or outflow, the equation should be modified to satisfy continuity. For the case in Fig. 3.7, we have

$$Q_3 = Q_1 + Q_2, \quad A_3 v_3 = A_1 v_1 + A_2 v_2 \tag{3.25}$$

$$Q_4 + Q_5 = Q_1 + Q_2, \quad A_4 v_4 + A_5 v_5 = A_1 v_1 + A_2 v_2 \tag{3.26}$$

Example 3.3 A three-dimensional incompressible flow field, known that $u_x = x^2 + z^2 + 5$ and $u_y = y^2 + z^2 - 3$, try to find u_z.
Solution
According to Eq. (3.20), we have

$$\frac{\partial u_z}{\partial z} = -\left(\frac{\partial u_x}{\partial x} + \frac{\partial u_y}{\partial y}\right) = -2(x+y)$$

By integration, we obtain

$$\int \frac{\partial u_z}{\partial z} dz = \int -2(x+y)dz$$
$$u_z = -2(x+y)z + C,$$

where C can be a constant or a function $f(x,y)$, thus:

$$u_z = -(x+y)z + f(x,y)$$

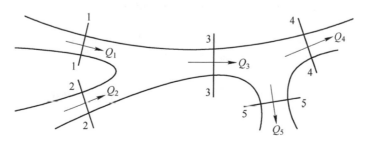

Fig. 3.7 Inflow and outflow

Fig. 3.8 Cyclonic separator

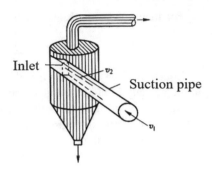

Example 3.4 Figure 3.8 is a cyclonic separator of dust. Its inlet is a rectangle with area $A_2 = 100$ mm $\times$ 20 mm, and the cross section of suction pipe is a circle with a diameter of 100 mm. The inlet velocity is $v_2 = 12$ m/s. Find v_1.

Solution

According to the continuity equation, we have

$$A_1 v_1 = A_2 v_2$$

so

$$v_1 = \frac{A_2 v_2}{A_1} = \frac{0.1 \times 0.02 \times 12}{\frac{\pi}{4} \times 0.1^2} = 3.06 \text{ m/s}$$

3.4   Differential Equations of Motion for Inviscid Fluid

This section focuses on the relationship between inviscid flow and force. For ideal fluids in motion, there is no viscosity and no inner friction generated, thus there are only mass force and pressure exerting on the element.

Take a cubic element in an ideal fluid in motion, as shown in Fig. 3.9. Its dimensions are δx, δy, δz. Assuming that the coordinates of the cubic element centroid are x, y, z, and its pressure is p. The velocity of the centroid is **u** with its components u_x, u_y, u_z. The fluid density is ρ, and the mass force is **J** with components X, Y, Z.

Since the pressure is a function of space and time, so the surface force and the mass force in the x-direction equal respectively

$$\left(p - \frac{\partial p}{\partial x} \frac{\delta x}{2}\right) \delta y \delta z - \left(p + \frac{\partial p}{\partial x} \frac{\delta x}{2}\right) \delta y \delta z \quad \text{and} \quad X \rho \delta x \delta y \delta z$$

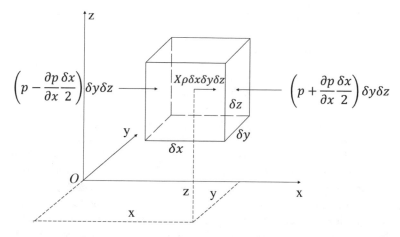

Fig. 3.9 Parallelepiped element [5]

According to Newton's second law, the sum of the surface force and mass force in the x axis equals the product of the element mass and the acceleration in the x axis:

$$\left(p - \frac{\partial p}{\partial x}\frac{\delta x}{2}\right)\delta y \delta z - \left(p + \frac{\partial p}{\partial x}\frac{\delta x}{2}\right)\delta y \delta z + X \rho \delta x \delta y \delta z = \rho \delta x \delta y \delta z \frac{du_x}{dt}$$

Simplifying the above equation, we can obtain the differential equation of motion in the x axis

Similarly

$$\left. \begin{array}{l} X - \frac{1}{\rho}\frac{\partial p}{\partial x} = \frac{du_x}{dt} \\[2mm] Y - \frac{1}{\rho}\frac{\partial p}{\partial y} = \frac{du_y}{dt} \\[2mm] Z - \frac{1}{\rho}\frac{\partial p}{\partial z} = \frac{du_z}{dt} \end{array} \right\} \tag{3.27}$$

Written in vectorial form, we have

$$\mathbf{J} - \frac{1}{\rho}\nabla p = \frac{d\mathbf{u}}{dt} \tag{3.28}$$

The above equation is the differential equation of motion for inviscid fluids, and was put forward first by Euler in 1755. Thus it is also called Eulerian differential equation of motion which is the basis of the classic fluid mechanics. For a static fluid with $u_x = u_y = u_z = 0$, the Eq. (3.27) can be transformed into Eulerian equilibrium differential Eq. (2.2). In other words, Eulerian equilibrium equation is a special case of Eulerian differential equation of motion.

Substituting the formula of acceleration into Eq. (3.27), we have

$$
\left.
\begin{array}{l}
X - \dfrac{1}{\rho}\dfrac{\partial p}{\partial x} = \dfrac{\partial u_x}{\partial x} u_x + \dfrac{\partial u_x}{\partial y} u_y + \dfrac{\partial u_x}{\partial z} u_z + \dfrac{\partial u_x}{\partial t} \\[2mm]
Y - \dfrac{1}{\rho}\dfrac{\partial p}{\partial y} = \dfrac{\partial u_y}{\partial x} u_x + \dfrac{\partial u_y}{\partial y} u_y + \dfrac{\partial u_y}{\partial z} u_z + \dfrac{\partial u_y}{\partial t} \\[2mm]
Z - \dfrac{1}{\rho}\dfrac{\partial p}{\partial z} = \dfrac{\partial u_z}{\partial x} u_x + \dfrac{\partial u_z}{\partial y} u_y + \dfrac{\partial u_z}{\partial z} u_z + \dfrac{\partial u_z}{\partial t}
\end{array}
\right\}
\qquad (3.29)
$$

The first three terms on the right-hand side of the above equations represent the change rate of velocity due to position change, which is called space-varying acceleration. The last term represents the change rate of velocity with respect to time, which is called time-varying acceleration. Therefore, the acceleration of fluid particle is the sum of space-varying acceleration and time-varying acceleration.

Generally speaking, there are four unknowns u_x, u_y, u_z and p in Eulerian differential equations of motion, but the equation number in (3.29) is only three. Combining with continuity equation, theoretically, these equations provide the possibility to solve the four unknowns. For inviscid fluid motion, the equations can be completely solved, but for viscous fluid motion, the solution is difficult to obtain due to the mathematical difficulty. Therefore, we can only get the integral or solution in a few special cases.

3.5 Bernoulli's Integral of Motion Differential Equations for Inviscid Fluid

In this section, we will discuss the integral of motion differential equations for inviscid fluid in special conditions, which is called Bernoulli's integral.

This integral is derived from the following conditions:

(1) Mass force is constant and potential

$$
X = \frac{\partial W}{\partial x}, Y = \frac{\partial W}{\partial y}, Z = \frac{\partial W}{\partial z}
$$

So total differential of the potential function $W = f(x, y, z)$ is

$$
\mathrm{d}W = \frac{\partial W}{\partial x}\,\mathrm{d}x + \frac{\partial W}{\partial y}\,\mathrm{d}y + \frac{\partial W}{\partial z}\,\mathrm{d}z = X\mathrm{d}x + Y\mathrm{d}y + Z\mathrm{d}z
$$

(2) The fluid is incompressible with constant ρ.
(3) The fluid flow is steady

$$
\frac{\partial p}{\partial t} = 0, \frac{\partial u_x}{\partial t} = \frac{\partial u_y}{\partial t} = \frac{\partial u_z}{\partial t} = 0
$$

Thus, the streamline and path line coincide, which means that streamline satisfies

$$\left.\begin{array}{l} dx = u_x dt \\ dy = u_y dt \\ dz = u_z dt \end{array}\right\}$$

Under these conditions above, multiplying the three equations in (3.27) by dx, dy and dz correspondingly and add together, we obtain

$$(X dx + Y dy + z dz) - \frac{1}{\rho}\left(\frac{\partial p}{\partial x} dx + \frac{\partial p}{\partial y} dy + \frac{\partial p}{\partial z} dz\right) = \frac{du_x}{dt} dx + \frac{du_y}{dt} dy + \frac{du_z}{dt} dz$$

According to integral conditions, we have

$$dW - \frac{1}{\rho} dp = \frac{du_x}{dt} u_x dt + \frac{du_y}{dt} u_y dt + \frac{du_z}{dt} u_z dt = u_x du_x + u_y du_y + u_z du_z$$

$$dW - \frac{1}{\rho} dp = d\left(\frac{u_x^2 + u_y^2 + u_z^2}{2}\right) = d\left(\frac{u^2}{2}\right)$$

Since ρ is constant, the above equation can be rewritten as

$$d\left(W - \frac{p}{\rho} - \frac{u^2}{2}\right) = 0$$

Integrating the above equation along streamline, we obtain

$$W - \frac{p}{\rho} - \frac{u^2}{2} = C \tag{3.30}$$

This is Bernoulli's integral of motion differential equations for inviscid fluid. It indicates that the value of the function $W - \frac{p}{\rho} - \frac{u^2}{2}$ keeps constant along the streamline when inviscid and incompressible fluid flows steadily under the action of potential mass force. In other words, the value of the function $W - \frac{p}{\rho} - \frac{u^2}{2}$ is invariable on a same streamline, while varies with different streamlines. As shown in Fig. 3.10, point 1 and 2 are on the same streamline, and we have

$$W_1 - \frac{p_1}{\rho} - \frac{u_1^2}{2} = W_2 - \frac{p_2}{\rho} - \frac{u_2^2}{2} \tag{3.31}$$

Fig. 3.10 Bernoulli's integral

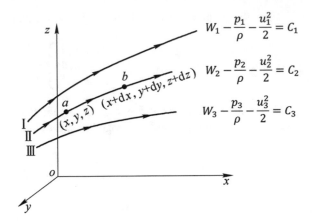

In general, there are many different kinds of mass force exerting on the flowing fluid, such as inertial force and gravity. But in many engineering problems, there is only gravity as mass force of which the components are

$$X = 0, \quad Y = 0, \quad Z = -g$$

So

$$dW = -gdz$$

By integrating

$$W = -gz + C$$

Substituting it into Eq. (3.30), we obtain

$$gz + \frac{p}{\rho} + \frac{u^2}{2} = \text{constant}$$

Dividing each term of the above equation by g and combining with $\gamma = \rho g$, we obtain

$$z + \frac{p}{\gamma} + \frac{u^2}{2g} = \text{constant} \tag{3.32}$$

Applying the above equation to two arbitrary points on the same streamline, we have

$$z_1 + \frac{p_1}{\gamma} + \frac{u_1^2}{2g} = z_2 + \frac{p_2}{\gamma} + \frac{u_2^2}{2g} \tag{3.33}$$

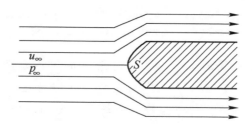

Fig. 3.11 Flow around an object

The above equation is called Bernoulli equation for incompressible and inviscid fluid. Since cross section area of the elementary flow beam is infinitesimal, the flow properties z, p and u are assumed to be the same. Thus Eqs. (3.32) and (3.33) can be applied to elementary flow beam, and then can be called Bernoulli equation along elementary flow beam of incompressible and inviscid fluid.

Example 3.5 A flow around an object is shown in Fig. 3.11. The velocity and pressure at the infinity upstream are respectively $u_\infty = 4.2$ m/s and $p_\infty = 0$. The water velocity declines due to the resistance of the object, and the velocity of the stagnation point S equals zero. Try to determine the pressure in the point S.
Solution
Assume the pressure of the stagnation point S is p_S, and the viscosity is ignored. According to Eq. (3.33), we have

$$z_\infty + \frac{p_\infty}{\gamma} + \frac{u_\infty^2}{2g} = z_S + \frac{p_S}{\gamma} + \frac{u_S^2}{2g}$$

Since $z_\infty = z_S$

$$\frac{p_S}{\gamma} = \frac{p_\infty}{\gamma} + \frac{u_\infty^2}{2g} - \frac{u_S^2}{2g} = \frac{4.2^2}{2 \times 9.8} = 0.9 \, \text{m}$$

So the pressure is

$$p_S = 0.9 \, \text{m} \, H_2O$$

3.6 Differential Equations of Motion and Bernoulli Equation for Viscous Fluid

3.6.1 Differential Equation of Motion for Viscous Fluid

Differential equation of motion for viscous fluid can also be obtained in terms of derivation method of inviscid fluid. The differential equations of motion for incompressible viscous fluid are given in (3.34) directly without derivation process, which are called Navier–Stokes equations (N-S equations).

$$
\left.
\begin{array}{l}
X - \frac{1}{\rho}\frac{\partial p}{\partial x} + v\nabla^2 u_x = \frac{\mathrm{d}u_x}{\mathrm{d}t} = \frac{\partial u_x}{\partial t} + u_x\frac{\partial u_x}{\partial x} + u_y\frac{\partial u_x}{\partial y} + u_z\frac{\partial u_x}{\partial z} \\[2mm]
Y - \frac{1}{\rho}\frac{\partial p}{\partial y} + v\nabla^2 u_y = \frac{\mathrm{d}u_y}{\mathrm{d}t} = \frac{\partial u_y}{\partial t} + u_x\frac{\partial u_y}{\partial x} + u_y\frac{\partial u_y}{\partial y} + u_z\frac{\partial u_y}{\partial z} \\[2mm]
Z - \frac{1}{\rho}\frac{\partial p}{\partial z} + v\nabla^2 u_z = \frac{\mathrm{d}u_z}{\mathrm{d}t} = \frac{\partial u_z}{\partial t} + u_x\frac{\partial u_z}{\partial x} + u_y\frac{\partial u_z}{\partial y} + u_z\frac{\partial u_z}{\partial z}
\end{array}
\right\},
\qquad (3.34)
$$

where ∇^2 is Laplace operator, $\nabla^2 = \frac{\partial^2}{\partial x^2} + \frac{\partial^2}{\partial y^2} + \frac{\partial^2}{\partial z^2}$.

Compared with Eulerian motion differential equations for ideal fluid, the viscosity term $v\nabla^2 u$ is added to N-S equations, so it is a complicated nonlinear partial differential equation. N-S equations combining continuity equation are theoretically enough to solve the four unknowns u_x, u_y, u_z and p. However, the solutions are very difficult to obtain for actual flow problem due to the complicated boundary conditions. With the development of the computer and computing, now there are several numerical solution methods for N-S equations.

3.6.2   Bernoulli Equation for Viscous Fluid Motion

Similar to the previous section, we only discuss the integral of motion differential equation for viscous fluid with the potential mass force. Equation (3.34) can be transformed into

$$
\left.
\begin{array}{l}
\frac{\partial}{\partial x}\left(W - \frac{p}{\rho} - \frac{u^2}{2}\right) + v\nabla^2 u_x = 0 \\[2mm]
\frac{\partial}{\partial y}\left(W - \frac{p}{\rho} - \frac{u^2}{2}\right) + v\nabla^2 u_y = 0 \\[2mm]
\frac{\partial}{\partial z}\left(W - \frac{p}{\rho} - \frac{u^2}{2}\right) + v\nabla^2 u_z = 0
\end{array}
\right\}
\qquad (3.35)
$$

If the flow is steady, the components of the fluid particle's infinitesimal displacement $\mathrm{d}l$ along the streamline are $\mathrm{d}x, \mathrm{d}y, \mathrm{d}z$. Multiplying the equations in (3.35) by $\mathrm{d}x, \mathrm{d}y, \mathrm{d}z$ correspondingly and adding together, we have

$$
\mathrm{d}\left(W - \frac{p}{\rho} - \frac{u^2}{2}\right) + v\left(\nabla^2 u_x\mathrm{d}x + \nabla^2 u_y\mathrm{d}y + \nabla^2 u_z\mathrm{d}z\right) = 0 \qquad (3.36)
$$

$v\nabla^2 u_x, v\nabla^2 u_y, v\nabla^2 u_z$ are the projections of the tangential stress/shear stress acting on per unit of viscous fluid, so the second term in the above equation is the work by the shear stress along the infinitesimal displacement $\mathrm{d}l$ on the streamline. The direction of the resultant force of these shear stresses is always opposite to fluid flow direction, so the work is negative. Thus, the second term in the above equation can be rewritten as $v\left(\nabla^2 u_x\mathrm{d}x + \nabla^2 u_y\mathrm{d}y + \nabla^2 u_z\mathrm{d}z\right) = -\mathrm{d}w_R$, where w_R is resistance work. Substituting it into Eq. (3.36), we obtain

$$d\left(W - \frac{p}{\rho} - \frac{u^2}{2} - w_R\right) = 0$$

Integrating the above equation along the streamline, we obtain

$$W - \frac{p}{\rho} - \frac{u^2}{2} - w_R = \text{constant} \tag{3.37}$$

This is Bernoulli's integral of motion differential equation for viscous fluid [6]. It indicates that the value of the function $W - \frac{p}{\rho} - \frac{u^2}{2} - w_R$ remains the same along the streamline when viscous and incompressible fluid flows steadily under the potential mass force. Taking arbitrarily point 1 and point 2 on the same streamline, we have

$$W_1 - \frac{p_1}{\rho} - \frac{u_1^2}{2} - w_{R1} = W_2 - \frac{p_2}{\rho} - \frac{u_2^2}{2} - w_{R2} \tag{3.38}$$

Assuming only gravity as mass force, the z axis is upward vertically, then we have

$$W_1 = -gz_1, W_2 = -gz_2$$

Substituting it into Eq. (3.38) and simplifying, we obtain

$$z_1 + \frac{p_1}{\gamma} + \frac{u_1^2}{2g} = z_2 + \frac{p_2}{\gamma} + \frac{u_2^2}{2g} + \frac{1}{g}(w_{R2} - w_{R1}), \tag{3.39}$$

where $w_{R2} - w_{R1}$ represents the increment of the work by inner friction when per unit of viscous fluid flows from point 1 to point 2. Using $h_l' = \frac{1}{g}(w_{R2} - w_{R1})$ to represent the resistance work, the Eq. (3.39) can be rewritten as

$$z_1 + \frac{p_1}{\gamma} + \frac{u_1^2}{2g} = z_2 + \frac{p_2}{\gamma} + \frac{u_2^2}{2g} + h_l' \tag{3.40}$$

This is Bernoulli equation for viscous fluid motion. It indicates that the value of $z + \frac{p}{\gamma} + \frac{u^2}{2g}$ decreases along the flow direction for viscous fluid flow. The above equation can be applied to elementary flow beam, then it can be called Bernoulli equation for elementary flow beam of viscous fluid.

3.6.3   Energy Considerations About Bernoulli Equation

Similar to fluid statics, $z, \frac{p}{\gamma}, \frac{u^2}{2g}$ are called elevation head, pressure head and velocity head respectively. h_l' is called head loss.

We also interpret the three terms $z, \frac{p}{\gamma}, \frac{u^2}{2g}$ as potential, "pressure" and kinetic energies to make up the total mechanical energy of the fluid in units of meters. h_l' is mechanical energy loss. We also define the total mechanical energy $H = z + \frac{p}{\gamma} + \frac{u^2}{2g}$ as energy grade line (EGL), and $z + \frac{p}{\gamma}$ as hydraulic grade line (HGL).

As shown in Fig. 3.12, the Bernoulli equation of inviscid fluid flow states that the total mechanical energy of the fluid along streamline keeps unchanged, namely, the EGL is the same. Therefore, the Bernoulli equation is a special representation of energy conservation law in the fluid mechanics. The Bernoulli equation of viscous fluid flow states that the EGL will continuously decrease in the direction of flow due to the mechanical energy loss.

Example 3.6 A Pitot tube with a mercury manometer is connected to a pipe. The diameter of the pipe is $D = 150$ mm, as shown in Fig. 3.13. The average velocity in the pipe is 0.84 times that on the axis of the pipe. Point 1 and 2 are very close to each other. Neglecting the resistance force in flow, try to determine the flow rate.
Solution
Neglect the energy loss when water flows from point 1 to 2, and establish Bernoulli equation for cross sections 1-1 and 2-2 (select the centerline of the pipe as datum reference):

$$z_1 + \frac{p_1}{\gamma_W} + \frac{u_1^2}{2g} = z_2 + \frac{p_2}{\gamma_W} + \frac{u_2^2}{2g}$$

Since $z_1 = z_2 = 0$, $u_1 = u_{\max}$ and $u_2 = 0$, we have

$$u_{\max} = \sqrt{2g \times \frac{p_2 - p_1}{\gamma_W}} \qquad (1)$$

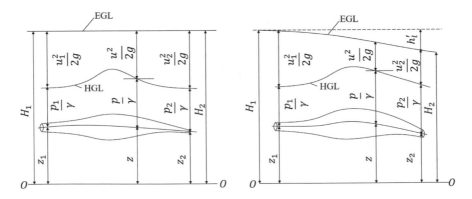

Fig. 3.12 EGL and HGL for inviscid and viscous fluid

Fig. 3.13 Pitot tube

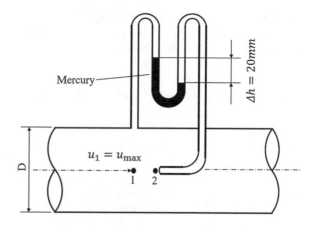

According to fluid statics

$$p_2 - p_1 = (\gamma_M - \gamma_W)\Delta h$$

Namely

$$\frac{p_2 - p_1}{\gamma_W} = \frac{(\gamma_M - \gamma_W)\Delta h}{\gamma_W}$$

Substituting it into Eq. (1), we obtain

$$u_{max} = \sqrt{2g \times \frac{(\gamma_M - \gamma_W)\Delta h}{\gamma_W}} = \sqrt{2 \times 9.8 \times \frac{(133280 - 9800) \times 0.02}{9800}} = 2.22\,\text{m/s}$$

Thus

$$v = 0.84u_{max} = 0.84 \times 2.22 = 1.87\,\text{m/s}$$

$$Q = Av = \frac{\pi \times 0.15^2}{4} \times 1.87 = 0.033\,\text{m}^3/\text{s} = 33\,\text{L/s}$$

3.7 Bernoulli Equation for Viscous Fluid Flow in Pipes and Ducts

To solve actual flow problems utilizing Bernoulli equation in engineering, Bernoulli equation for elementary flow beam should be extended to flow in pipes and ducts. The properties such as density, velocity and pressure, etc., on cross section vary with streamlines when fluid flows in a pipe or duct. The Bernoulli equation along a streamline cannot be applied in a pipe or duct directly. Therefore, we have to derive the Bernoulli equation for fluid flow in pipes and ducts.

3.7.1 Rapidly Varied Flow and Gradually Varied Flow

Figure 3.14 is the sketch of rapidly and gradually varied flow. The feature of rapidly varied flow is the large angle β between streamlines and the small curvature radius r of the streamline. As shown in Fig. 3.14, flows in sections 1-2, 2-3, and 4-5 are rapidly varied flow. Inertial force cannot be neglected and the component of inner friction on the cross section is not equal to 0 for rapidly varied flow. There are some complicated forces on these cross sections, so it is unsuitable to apply Bernoulli equation between them.

The feature of gradually varied flow is the small angle between streamlines and the large curvature radius of the streamline. The streamlines are nearly parallel straight lines. As shown in Fig. 3.14, flows in sections 3-4 and 5-6 are gradually varied flow. In gradually varied flow, cross sections are almost planes. The projection of inner friction on the cross section is almost equal to 0. Thus, the pressure distribution on the cross section conforms to hydrostatic pressure distribution.

It can be proved that for every point of the same cross section in gradually varied flow, the relationship between its pressure and position can be written as

$$z + \frac{p}{\gamma} = \text{constant} \tag{3.41}$$

It is notable that the value of the constant in the above equation varies with cross section. It can be seen from the piezometer tubes fixed on the cross sections for gradually varied flow that the HGL is invariable on the same cross section.

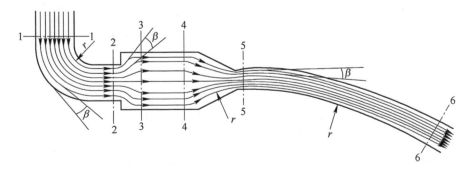

Fig. 3.14 Rapidly and gradually varied flow [5]

Fig. 3.15 Bernoulli equation for pipe flow

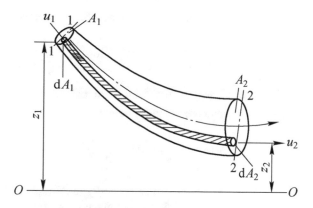

3.7.2 Bernoulli Equation in Pipe or Duct

For incompressible and viscous steady flow shown in Fig. 3.15, select an elementary flow beam, then its Bernoulli equation is

$$z_1 + \frac{p_1}{\gamma} + \frac{u_1^2}{2g} = z_2 + \frac{p_2}{\gamma} + \frac{u_2^2}{2g} + h_l'$$

Assuming that the fluid weight passing through this elementary flow beam per unit time is γdQ, we have

$$z_1 \gamma dQ + \frac{p_1}{\gamma} \gamma dQ + \frac{u_1^2}{2g} \gamma dQ = z_2 \gamma dQ + \frac{p_2}{\gamma} \gamma dQ + \frac{u_2^2}{2g} \gamma dQ + h_l' \gamma dQ$$

Integrating every term over flow rate along corresponding cross section, then the energy equation for pipe flow is

$$\int_Q z_1 \gamma dQ + \int_Q \frac{p_1}{\gamma} \gamma dQ + \int_Q \frac{u_1^2}{2g} \gamma dQ = \int_Q z_2 \gamma dQ + \int_Q \frac{p_2}{\gamma} \gamma dQ + \int_Q \frac{u_2^2}{2g} \gamma dQ + \int_Q h_l' \gamma dQ$$

$$(3.42)$$

The integral of above equation can be divided into three parts. The first part is the first two terms on each side of the equation, which can be written as

$$\int_Q z\gamma dQ + \int_Q \frac{p}{\gamma} \gamma dQ = \int_Q \left(z + \frac{p}{\gamma}\right) \gamma dQ = \gamma \int_A \left(z + \frac{p}{\gamma}\right) u dA$$

In the section of a gradually varied flow we have $z + \frac{p}{\gamma} = $ constant, thus

$$\gamma \int_A \left(z + \frac{p}{\gamma} \right) u \mathrm{d}A = \gamma \left(z + \frac{p}{\gamma} \right) \int_A u \mathrm{d}A = \left(z + \frac{p}{\gamma} \right) \gamma Q$$

The second part is the third term of the equation $\int_Q \frac{u_1^2}{2g} \gamma \mathrm{d}Q$, and it can be expressed by average velocity, namely

$$\int_Q \frac{u_1^2}{2g} \gamma \mathrm{d}Q = \int_A \frac{1}{2} \rho u^3 \mathrm{d}A = \int_A \frac{1}{2} (\rho u \mathrm{d}A) u^2 = \alpha \left(\frac{1}{2} \rho v^3 A \right) = \frac{\alpha v^2}{2g} \gamma Q,$$

where α is called the kinetic energy coefficient. According to measurement, $\alpha = 1.05$–1.10 in actual flows. We always use the approximation $\alpha = 1$ in our pipe flow calculations.

The third part is the last term of the equation $\int_Q h_l' \gamma \mathrm{d}Q$, which represents the mechanical energy loss when fluid particle flows from cross section 1-1 to cross section 2-2. Letting h_l denote the average energy loss per unit weight fluid, we obtain

$$\int_Q h_l' \gamma \mathrm{d}Q = h_l \gamma Q$$

Substituting the three parts into Eq. (3.42) and dividing every term by γQ, then the energy per unit weight fluid for pipe flow can be written as

$$z_1 + \frac{p_1}{\gamma} + \frac{\alpha_1 v_1^2}{2g} = z_2 + \frac{p_2}{\gamma} + \frac{\alpha_2 v_2^2}{2g} + h_l \qquad (3.43)$$

This is the Bernoulli equation for incompressible steady pipe flow under the action of gravity, which is important in engineering fluid mechanics. Equation (3.43) can be applied only to the following engineering applications

(1) The fluid is incompressible;
(2) The flow is steady;
(3) The mass force is only due to gravity;
(4) The flows on two cross sections belong to gradually varied flows, but there may be rapidly varied flows between the two cross sections;
(5) There is no inflow or outflow between the two cross sections and the flow process is adiabatic.

3.7.3 Other Forms of Bernoulli Equation

(1) Bernoulli equation for gas flow

Bernoulli Eq. (3.43) for steady flow is also suitable for incompressible gas flow, but for gas flow, its specific weight generally varies. Neglecting the effect of internal energy, then the Bernoulli equation for gas flow is

$$z_1 + \frac{p_1}{\gamma_1} + \frac{\alpha_1 v_1^2}{2g} = z_2 + \frac{p_2}{\gamma_2} + \frac{\alpha_2 v_2^2}{2g} + h_l \qquad (3.44)$$

(2) Bernoulli equation with energy input and output

If there is energy input or output between two cross sections, letting $\pm E$ denote the input or output energy, then Bernoulli equation is

$$z_1 + \frac{p_1}{\gamma} + \frac{\alpha_1 v_1^2}{2g} \pm E = z_2 + \frac{p_2}{\gamma} + \frac{\alpha_2 v_2^2}{2g} + h_l \qquad (3.45)$$

If fluid machinery does work on fluid, namely inputs energy to the system, the sign of E is positive, such as pump or fan. If the fluid does work on fluid machinery, namely system outputs energy, the sign of E is negative, for example, water turbine pipeline system.

(3) Bernoulli equation with inflow or outflow

If there is inflow between the two cross sections shown in Fig. 3.16a, then Bernoulli equation is

$$\left. \begin{array}{l} z_1 + \frac{p_1}{\gamma} + \frac{\alpha_1 v_1^2}{2g} = z_3 + \frac{p_3}{\gamma} + \frac{\alpha_3 v_3^2}{2g} + h_{l1-3} \\ z_2 + \frac{p_2}{\gamma} + \frac{\alpha_2 v_2^2}{2g} = z_3 + \frac{p_3}{\gamma} + \frac{\alpha_3 v_3^2}{2g} + h_{l2-3} \end{array} \right\} \qquad (3.46)$$

If there is outflow between the two cross sections shown in Fig. 3.16b, then Bernoulli equation is

$$\left. \begin{array}{l} z_1 + \frac{p_1}{\gamma} + \frac{\alpha_1 v_1^2}{2g} = z_2 + \frac{p_2}{\gamma} + \frac{\alpha_2 v_2^2}{2g} + h_{l1-2} \\ z_1 + \frac{p_1}{\gamma} + \frac{\alpha_1 v_1^2}{2g} = z_3 + \frac{p_3}{\gamma} + \frac{\alpha_3 v_3^2}{2g} + h_{l1-3} \end{array} \right\} \qquad (3.47)$$

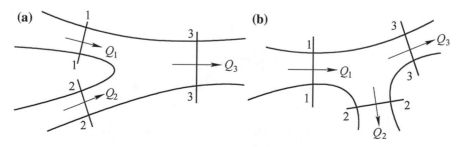

Fig. 3.16 Inflow and outflow

Continuity equations for the two cases are respectively
    Inflow: $Q_1 + Q_2 = Q_3$;
    Outflow: $Q_1 = Q_2 + Q_3$.

3.7.4 Application Examples of Bernoulli Equation

Example 3.7 There is a water supply pipeline *AB* shown in Fig. 3.17. The pipe
diameter is $D = 300\,mm$, and the flow rate is $Q = 0.04\,m^3/s$. The relative pressure
of point B is 9.8×10^4 Pa, and the Height is $H = 20\,m$. Try to determine the head
loss in pipeline *AB*.
Solution
Select *O-O* as datum reference, and establish Bernoulli equation for cross
section 1-1 and 2-2:

$$z_1 + \frac{p_1}{\gamma} + \frac{\alpha_1 v_1^2}{2g} = z_2 + \frac{p_2}{\gamma} + \frac{\alpha_2 v_2^2}{2g} + h_l$$

According to the given data

$$z_1 = H = 20\,m, \quad z_2 = 0$$

Then

$$\frac{p_1}{\gamma} = 0, \quad \frac{p_2}{\gamma} = \frac{1 \times 9.8 \times 10^4}{9800} = 10\,m\,H_2O, \quad \alpha_1 = \alpha_2 = 1, \quad v_1 \approx 0$$

$$v_2 = \frac{Q}{A} = \frac{0.04}{\frac{\pi}{4} \times 0.3^2} = 0.566\,m/s$$

Fig. 3.17 Water supply
pipeline

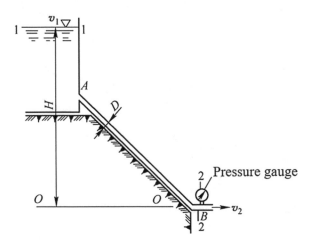

Substituting above values into Bernoulli equation, we obtain

$$h_l = z_1 + \frac{p_1}{\gamma} + \frac{\alpha_1 v_1^2}{2g} - z_2 - \frac{p_2}{\gamma} - \frac{\alpha_2 v_2^2}{2g} = 20 - 10 - \frac{0.566^2}{2 \times 9.8} = 9.98 \, \text{m} \, \text{H}_2\text{O}$$

Example 3.8 Figure 3.18 is the suction pipe of an axial fan. The pipe inner diameter is $D = 0.3$ m, and the specific weight of air is $\gamma_a = 12.6 \, \text{N/m}^3$, and $\Delta h = 0.2$ m. Try to determine the flow rate Q.

Solution

Since the suction pipe is not long, the energy loss can be ignored and the air is regarded as incompressible fluid. Select O-O as datum reference, and establish Bernoulli equation for cross section 1-1 and 2-2:

$$z_1 + \frac{p_1}{\gamma_a} + \frac{\alpha_1 v_1^2}{2g} = z_2 + \frac{p_2}{\gamma_a} + \frac{\alpha_2 v_2^2}{2g}$$

According to the given data

$$z_1 = z_2 = 0, \quad p_1 = p_A = p_a, \quad p_2 = p_B = p_C = p_a - \gamma_w \Delta h, \quad v_1 \approx 0$$

Thus

$$v_2 = \sqrt{2g \frac{p_1 - p_2}{\gamma_a}} = \sqrt{2g \frac{p_a - (p_a - \gamma_w \Delta h)}{\gamma_a}}$$

$$= \sqrt{2g \frac{\gamma_w \Delta h}{\gamma_a}} = \sqrt{2 \times 9.8 \times \frac{9800 \times 0.2}{12.6}}$$

$$= 55.2 \, \text{m/s}$$

Fig. 3.18 Suction pipe of an axial fan

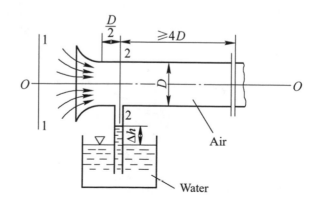

The flow rate is

$$Q = A_2 v_2 = \frac{\pi \times 0.3^2}{4} \times 55.2 = 3.90 \, \text{m}^3/\text{s}$$

Example 3.9 Figure 3.19 is a water pump piping system. The diameter of all pipes is 200 mm, and the flow rate is $Q = 0.06 \, \text{m}^3/\text{s}$. The height difference between reservoir C and reservoir A is $H = 25 \, \text{m}$. The head loss of the pipeline $A - B - C$ is $h_l = 5 \, \text{m}$, try to determine the input energy E from the pump to the system.
Solution

Select O-O as datum reference, and establish Bernoulli equation for cross section 1-1 and 2-2:

$$z_1 + \frac{p_1}{\gamma} + \frac{\alpha_1 v_1^2}{2g} + E = z_2 + \frac{p_2}{\gamma} + \frac{\alpha_2 v_2^2}{2g} + h_l$$

According to the given data

$$z_1 = 0, \quad z_2 = 25, \quad p_1 = p_2 = p_a, \quad v_1 = v_2 \approx 0, \quad h_l = 5 \, \text{m}$$

Thus

$$E = z_2 + h_l = 25 + 5 = 30 \, \text{m} \, H_2O$$

In engineering $E = H$ is called the head rise of the pump, which is used to raise water level and overcome the resistance in the pipeline.

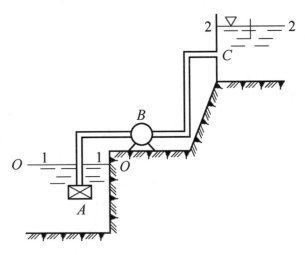

Fig. 3.19 Water pump piping system

3.8 Instruments for Velocity and Flow Rate Measurement

Instruments for velocity and flow rate measurement in engineering are invented in terms of Bernoulli equation. Here we mainly introduce two instruments of velocity and flow rate measurement respectively: Pitot tube and Venturi tube.

3.8.1 Pitot Tube

A Pitot tube, also known as Pitot probe, is a pressure measurement instrument used to measure fluid flow velocity.

A simple Pitot tube is a $90°$ glass tube with two open ends, as shown in Fig. 3.20. The procedure to measure the velocity of point M is as follow: put one end of the tube at point M, fluid flows into the pipe and its velocity decreases to 0 after reaching a certain height. Point M is called the stagnation point. There is another point M_0 close to point M and they are on the same streamline. The velocity of point M_0 is u. According to Bernoulli equation, we have

$$z_{M_0} + \frac{p_{M_0}}{\gamma} + \frac{u^2}{2g} = z_M + H = z_M + \frac{p_M}{\gamma} + h$$

$z_{M_0} = z_M$, since point M_0 and point M are very close to each other, so $p_{M_0} = p_M$, thus

$$u = \sqrt{2gh} \tag{3.48}$$

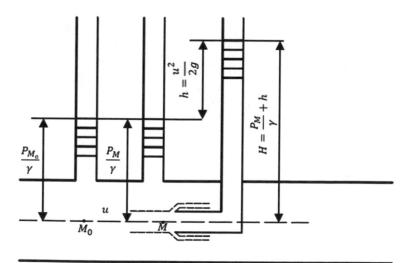

Fig. 3.20 Pitot tube

It shows that $\frac{u^2}{2g}$, the kinetic energy of point M_0, is transformed into the pressure energy h of the stagnation point M. However, the energy loss cannot be ignored since actual fluids are viscous. So the above equation can be corrected as

$$u = c\sqrt{2gh}, \tag{3.49}$$

where c is called velocity coefficient of Pitot tube, generally ranging from 0.97 to 0.99. For high-precision Pitot tubes, c can be approximately equal to 1.

3.8.2 Venturi Tube

A Venturi tube is used to measure the flow rate in the pipe, and it is composed of converging section A, constricted section (choke) B and diverging section C, as shown in Fig. 3.21.

Two piezometer tubes are fixed on the converging section and constricted section, respectively. For steady flow of inviscid fluid, selecting O-O as datum reference, establish Bernoulli equation between cross section 1-1 and 2-2:

$$z_1 + \frac{p_1}{\gamma} + \frac{v_1^2}{2g} = z_2 + \frac{p_2}{\gamma} + \frac{v_2^2}{2g}$$

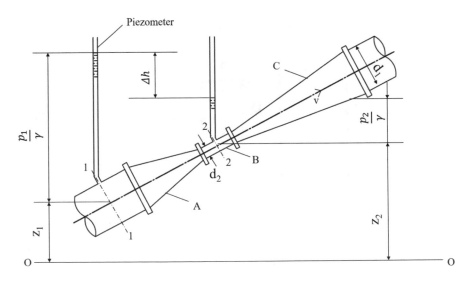

Fig. 3.21 Venturi tube [5]

From continuity equation we obtain

$$A_1 v_1 = A_2 v_2 \quad v_2 = \frac{A_1}{A_2} v_1 = \left(\frac{\pi d_1^2}{4} \Big/ \frac{\pi d_2^2}{4} \right) v_1 = \frac{d_1^2}{d_2^2} v_1$$

Substituting it into Bernoulli equation we obtain

$$\left(z_1 + \frac{p_1}{\gamma} \right) - \left(z_2 + \frac{p_2}{\gamma} \right) = \frac{v_1^2}{2g} \left(\frac{d_1^4}{d_2^4} - 1 \right)$$

$$v_1 = \frac{1}{\sqrt{\frac{d_1^4}{d_2^4} - 1}} \sqrt{2g \left[\left(z_1 + \frac{p_1}{\gamma} \right) - \left(z_2 + \frac{p_2}{\gamma} \right) \right]}$$

Assuming that $\dfrac{\sqrt{2g}}{\sqrt{\frac{d_1^4}{d_2^4} - 1}} = k$ and $\left(z_1 + \frac{p_1}{\gamma} \right) - \left(z_2 + \frac{p_2}{\gamma} \right) = \Delta h$, then

$$v_1 = k\sqrt{\Delta h}, \tag{3.50}$$

where k is called apparatus constant determined by its structure. Thus, the flow rate is

$$Q = A_1 v_1 = \frac{\pi d_1^2}{4} k\sqrt{\Delta h} \tag{3.51}$$

Since energy loss has been neglected, the above equation needs to be corrected as

$$Q = \mu \frac{\pi d_1^2}{4} k\sqrt{\Delta h}, \tag{3.52}$$

where μ is called flow rate coefficient of Venturi tube determined by its material, size, installation quality, fluid viscosity and so on. Its value can only be determined by experiments and usually ranges from 0.95 to 0.98.

To reduce energy loss, the inner wall of Venturi tube is often designed to be streamlined in engineering which is called Venturi nozzle. Venturi tube and Ventuti nozzle are widely utilized. However, the pressure in the constricted section should not be too low since it will result in vaporization and destroy the continuity of fluid flow, which causes Ventuti tube to malfunction.

Example 3.10 Figure 3.22 is a Venturi tube. Its diameters are $D = 100$ mm and $d = 50$ mm. The hydraulic grade lines (HGL) are $z_1 + \frac{p_1}{\gamma} = 1.0$ m and $z_2 + \frac{p_2}{\gamma} = 0.6$ m respectively. The flow rate coefficient is $\mu = 0.98$, try to determine the flow rate in the pipe.

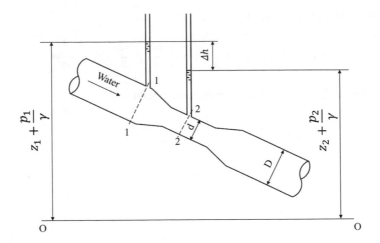

Fig. 3.22 Venturi tube

Solution

The height difference of hydraulic grade line between cross section 1-1 and 2-2 is

$$\Delta h = \left(z_1 + \frac{p_1}{\gamma} \right) - \left(z_2 + \frac{p_2}{\gamma} \right) = 1.0 - 0.6 = 0.4 \, \text{m}$$

According to Eq. (3.52), we have

$$Q = \mu \frac{\pi d_1^2}{4} k \sqrt{\Delta h} = \mu \frac{\pi d_1^2 \sqrt{2g}}{4\sqrt{\frac{d_1^4}{d_2^4} - 1}} \sqrt{\Delta h}$$

$$= 0.98 \times \frac{\pi \times 0.1^2 \sqrt{2 \times 9.8}}{4 \times \sqrt{(0.1/0.05)^4 - 1}} \sqrt{0.4}$$

$$= 0.00556 \, \text{m}^3/\text{s}$$

3.9 Momentum Equation for Steady Flow and Its Application

Fluid momentum equation is the concrete expression for the law of momentum conservation in fluid motion, which reflects the relationship between fluid momentum and force and is widely used to solve many fluid mechanics problems in engineering.

3.9.1   Momentum Equation for Steady Flow

According to Physics, momentum theorem states that the change rate of momentum with respect to time equals the resultant vector of external forces acting on the body,

$$\frac{d}{dt}\left(\sum m\mathbf{v}\right) = \frac{d\mathbf{M}}{dt} = \sum \mathbf{F} \tag{3.53}$$

Now apply this theorem to steady fluid flow. Select an elementary flow beam 1-2, and its cross sections are 1-1 and 2-2, as shown in Fig. 3.23. The pressure of section 1-1 and 2-2 are p_1 and p_2, respectively. The velocity of them are $\mathbf{v}_1$ and $\mathbf{v}_2$, respectively. After time dt, flow beam 1-2 has moved to 1'-2' along the streamline and its momentum has also changed.

The momentum change equals the difference between momentum $\mathbf{M}_{1'-2'}$ of flow beam 1'-2' and momentum $\mathbf{M}_{1-2}$ of flow beam 1-2. For steady flow, the momentum of flow beam 1'-2 keeps invariable, so the momentum change equals the difference between momentum of flow beam 2-2' and momentum of flow beam 1-1'

$$d\mathbf{M} = \mathbf{M}_{2-2'} - \mathbf{M}_{1-1'} = dm_2\mathbf{u}_2 - dm_1\mathbf{u}_1 = \rho dQ_2 dt\mathbf{u}_2 - \rho dQ_1 dt\mathbf{u}_1$$

Applying the above equation to the pipe flow, we have

$$\sum d\mathbf{M} = \int_{Q_2} \rho dQ_2 dt\mathbf{u}_2 - \int_{Q_1} \rho dQ_1 dt\mathbf{u}_1 = \rho dt\left(\int_{Q_2} dQ_2\mathbf{u}_2 - \int_{Q_1} dQ_1\mathbf{u}_1\right) \tag{3.54}$$

From continuity equation, we have

$$\int_{Q_2} dQ_2 = Q_2 = \int_{Q_1} dQ_1 = Q_1 = Q$$

Fig. 3.23 Momentum equation

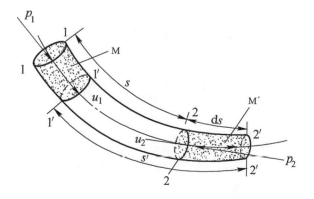

Substituting average velocity v into Eq. (3.54), we obtain

$$\sum d\mathbf{M} = \rho Q dt(\alpha_{02}\mathbf{v}_2 - \alpha_{01}\mathbf{v}_1),$$

where α is a coefficient, similar to kinetic energy coefficient.

From Eq. (3.53), we have

$$\sum \mathbf{F} = \rho Q(\alpha_{02}\mathbf{v}_2 - \alpha_{01}\mathbf{v}_1) \tag{3.55}$$

The above equation is the momentum equation for incompressible steady flow. $\sum \mathbf{F}$ is the resultant of the external force acting on the fluid, including the gravity $\mathbf{G}$ of flow beam 1-2, the pressure $\mathbf{P}_1$, $\mathbf{P}_2$ on two cross sections, and the summation $\mathbf{R}$ of surface forces acting on other surfaces, so the above equation can be written as

$$\sum \mathbf{F} = \mathbf{G} + \mathbf{P}_1 + \mathbf{P}_2 + \mathbf{R} = \rho Q(\alpha_{02}\mathbf{v}_2 - \alpha_{01}\mathbf{v}_1) \tag{3.56}$$

In general engineering calculation $\alpha_{02} = \alpha_{01} = 1$, and component form of momentum equation can be written as

$$\left.\begin{array}{l} \sum F_x = \rho Q(v_{2x} - v_{1x}) \\ \sum F_y = \rho Q(v_{2y} - v_{1y}) \\ \sum F_z = \rho Q(v_{2z} - v_{1z}) \end{array}\right\} \tag{3.57}$$

Momentum equation is of significant importance and usually used to determine the interaction force between the fluid and solid surface.

3.9.2 Application of Momentum Equation

(1) Force acting on the tube wall by fluid

Figure 3.24a is a converging elbow, the average velocities of cross section 1-1 and 2-2 are v_1 and v_2, respectively. The forces acting on the fluid between cross section 1-1 and 2-2 (Fig. 3.24b) include fluid gravity G, the force R by the elbow, forces p_1A_1 and p_2A_2 on cross sections. Establishing coordinate system as shown in the figure, the momentum equations in x and z directions are

$$\left.\begin{array}{l} \sum F_x = p_1A_1 - p_2A_2 \cos\theta - R_x = \rho Q(v_{2x} - v_{1x}) \\ \sum F_z = -p_2A_2 \sin\theta - G + R_z = \rho Q(v_{2z} - v_{1z}) \end{array}\right\}$$

So

$$\left.\begin{array}{l} R_x = p_1A_1 - p_2A_2 \cos\theta - \rho Q(v_2 \cos\theta - v_1) \\ R_z = p_2A_2 \sin\theta + G + \rho Q v_2 \sin\theta \end{array}\right\} \tag{3.58}$$

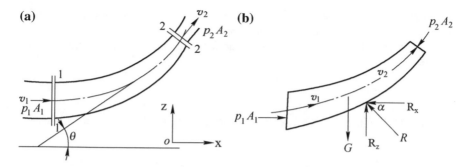

Fig. 3.24 Force acting on an elbow by fluid

The magnitude and direction of the resultant force are $R = \sqrt{R_x^2 + R_z^2}$ and $\alpha = \text{arctg}\frac{R_z}{R_x}$, respectively.

The force F acting on the elbow by the fluid equals R, but opposite in direction.

Especially, for variable radius elbow when $\theta = 90°$ and $Q = A_1 v_1 = A_2 v_2$, the force acting on elbow by the fluid is

$$\left.\begin{array}{l} F_x = (p_1 + \rho v_1^2)A_1 \\ F_z = (p_2 + \rho v_2^2)A_2 + G \end{array}\right\} \tag{3.59}$$

For equal radius elbow when $\theta = 90°$ and $A_1 = A_2 = A$, if the elbow is on a horizontal level then the force acting on elbow by the fluid is

$$\left.\begin{array}{l} F_x = (p_1 + \rho v^2)A \\ F_z = (p_2 + \rho v^2)A \end{array}\right\} \tag{3.60}$$

(2) Forces acting on the baffle by the jet

A horizontal jet rushes to a baffle with inclined angle θ, as shown in Fig. 3.25. The cross-section area and average velocity of the jet are A_0 and v_0 respectively. The jet

Fig. 3.25 A jet rushed on the baffle

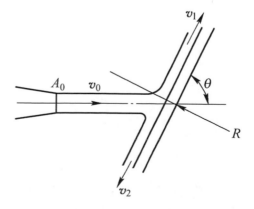

is divided into two streams with v_1 and v_2, respectively. Take the jet as control volume. The force acting on the jet by the baffle in its normal direction is R. Neglecting diffusion, resistance and head loss, according to Bernoulli equation, we have

$$v_1 = v_2 = v_0$$

Take the baffle as x axis and its normal direction as y axis, the momentum equations are

$$\left. \begin{array}{l} \sum F_x = 0 = \rho(Q_1 v_1 - Q_2 v_2 - Q_0 v_0 \cos \theta) \\ \sum F_y = -R = -\rho Q_0 v_0 \sin \theta \end{array} \right\} \tag{3.61}$$

From continuity equation, we have $Q_1 + Q_2 = Q_0$, then

$$\left. \begin{array}{l} Q_1 = \frac{Q_0}{2}(1 + \cos \theta); Q_2 = \frac{Q_0}{2}(1 - \cos \theta) \\ R = \rho Q_0 v_0 \sin \theta = \rho A_0 v_0^2 \sin \theta \end{array} \right\} \tag{3.62}$$

The force acting on the baffle by the jet is F, which equals R, but opposite in direction. When $\theta = 90°$, the jet rushes along the normal direction of the baffle, we have

$$\left. \begin{array}{l} Q_1 = Q_2 = \frac{Q_0}{2} \\ R = \rho A_0 v_0^2 \end{array} \right\} \tag{3.63}$$

(3)   Reverse thrust of the jet

An open container is filled with liquid, and there is a small hole on the container's side wall. The area of the hole is A, as shown in Fig. 3.26. Assuming that the flow rate is very small, so it can be regarded as steady flow for a short time,

Fig. 3.26 Reverse thrust of the jet

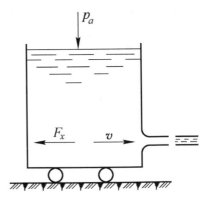

namely its velocity is $v = \sqrt{2gh}$. Thus, the change rate of the momentum in the x-direction is

$$\frac{\mathrm{d}}{M}\mathrm{d}t = \rho Q v = \rho A v^2$$

According to momentum theorem, this is the projection of the force acting on the fluid by the container on the x axis, namely $R_x = \rho A v^2$, and the reverse thrust of the jet is $F_x = -R_x = -\rho A v^2$. If the container can move freely along the x axis, then it will move in the opposite direction of the jet driven by F_x, which is the reverse thrust of the jet. Rockets, jet aircrafts are all driven by this kind of reverse thrust.

Example 3.11 There is a horizontal pipe with a nozzle fixed at its right end, and their diameters are respectively $D = 100\,\text{mm}$ and $d = 50\,\text{mm}$, as shown in Fig. 3.27. The flow rate in the pipe is $Q = 1\,\text{m}^3/\text{min}$, try to determine the component on x axis of the force acting on the jet by the nozzle.
Solution
From continuity equation, we have

$$v_1 = \frac{Q}{A_1} = \frac{Q}{\frac{\pi D^2}{4}} = \frac{\frac{1}{60} \times 4}{\pi \times 0.1^2} = 2.123\,\text{m/s}$$

$$v_2 = \frac{Q}{A_2} = \frac{Q}{\frac{\pi d^2}{4}} = \frac{\frac{1}{60} \times 4}{\pi \times 0.05^2} = 8.492\,\text{m/s}$$

Select O-O as datum reference, and establish Bernoulli equation for cross section 1-1 and 2-2

$$z_1 + \frac{p_1}{\gamma} + \frac{v_1^2}{2g} = z_2 + \frac{p_2}{\gamma} + \frac{v_2^2}{2g}$$

Fig. 3.27 Nozzle

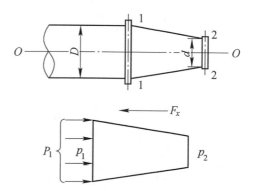

Since $z_1 = z_2$ and $p_2 = 0$, so

$$p_1 = \frac{\gamma}{2g}(v_2^2 - v_1^2) = \frac{9800}{2 \times 9.8}(8.496^2 - 2.123^2) = 33837\,\text{N/m}^2,$$

Assuming that the component on x axis of the force acting on the jet by the nozzle is F_x, the momentum equation of the jet is

$$p_1A_1 - F_x = \rho Q(v_2 - v_1)$$

Thus

$$F_x = p_1A_1 - \rho Q(v_2 - v_1) = 33837 \times \frac{\pi}{4} \times 0.1^2 - 1000 \times \frac{1}{60}(8.496 - 2.123)$$

$$= 159.4\,\text{N}$$

Example 3.12 Figure 3.28 is a nozzle with a 180° arc slit. The velocity of the jet v is 15 m/s, and its thickness t is 0.03 m. Other dimensions are $D = 0.2\,\text{m}$ and $R = 0.3\,\text{m}$, try to determine

(1) the flow rate of the jet;
(2) the component of the force needed to keep the arc nozzle still in the y direction.

Solution
From continuity equation, we have

$$Q_v = \pi R t v = 3.14 \times 0.3 \times 0.03 \times 15 = 0.424\,\text{m}^3/\text{s}$$

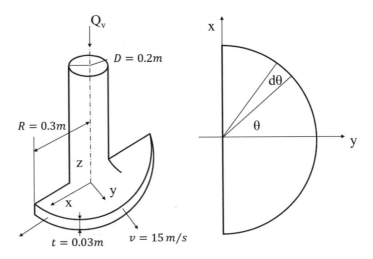

Fig. 3.28 Jet nozzle

The momentum equation in the y direction is

$$F_y = 2 \int_0^{\frac{\pi}{2}} \rho v dQ = 2 \int_0^{\frac{\pi}{2}} \rho v \cos \theta R t v d\theta = 2\rho v^2 R t = 4.05 \text{ kN}$$

3.10 Problems

3.1 A flow field is given by $u_x = 2kx$, $u_y = 2ky$ and $u_z = -4kz$ (k is constant). Try to determine the streamline through point (1, 0, 1).

3.2 A flow field is given by $u_x = 1 + At$ and $u_y = 2x$ (A is constant). Try to determine the streamline through point (x_0, y_0) at $t = t_0$.

3.3 A flow field is $\mathbf{u} = (6 + x^2y + t^2)\mathbf{i} - (xy^2 + 10t)\mathbf{j} + 25\mathbf{k}$, try to determine the acceleration in point (3, 2, 0) at $t = 1$.

3.4 A flow field for incompressible fluid is

$$\begin{cases} u_x = xt + 2y \\ u_y = xt^2 - yt \end{cases}$$

Try to determine the acceleration in point $A(1, 2)$ at $t = 1$ s.

3.5 As shown in Fig. 3.29, the diameters of the two pipes are $d_1 = 5$ m and $d_2 = 1$ m, respectively. The velocity distribution on the cross section for the big diameter pipe is $u = 6.25 - r^2 \text{ m/s}$, where r denotes the radius. Try to determine the flow rate and the average velocity of the smaller pipe.

3.6 The velocity distribution on the cross section of the pipe is $u = u_{max}\left[1 - \left(\frac{r}{r_0}\right)^2\right]$, where u_{max} is the maximum velocity on the axis of the pipe, r_0 is the radius of the pipe, and r is the distance between the point and the axis. Try to determine the average velocity of the cross section.

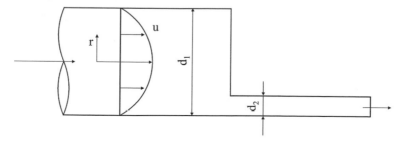

Fig. 3.29 Problem 3.5

3.7   For three-dimensional incompressible flow, it is known the velocity field is
$u_x = x^2 + y^2 z^3$, $u_y = -(xy + yz + zx)$, and $u_z = 0$ when $z = 0$. Try to determine the expression of u_z.

3.8   As shown in Fig. 3.30, there is an outflow in the pipeline. The dimensions are
$d_A = 45$ cm, $d_B = 30$ cm, $d_C = 20$ cm, $d_D = 15$ cm. $v_A = 2$ m/s   and
$v_C = 4$ m/s. Try to determine v_B and v_D.

3.9   Figure 3.31 is a steam pipeline. The dimensions of the pipeline are
$d_0 = 50$ mm, $d_1 = 45$ mm, $d_2 = 40$ mm. The velocity at the inlet is
$v_0 = 25$ m/s. The densities for different parts are $\rho_0 = 2.62 \, \text{kg/m}^3$, $\rho_1 =$
$2.24 \, \text{kg/m}^3$ and $\rho_2 = 2.30 \, \text{kg/m}^3$ respectively. Try to determine the average
velocity v_1 and v_2 to make sure that the flow rate of the two right pipes equals
each other.

3.10   Figure 3.32 is a drain pipe with eight identical orifices. The diameters of the
pipe and orifice are $D = 2$ cm and $d = 1$ mm, respectively. The liquid flows
into the pipe with average velocity $v = 0.15$ m/s, and the outflow velocity of
each orifice decreases by 2% than the former. Try to determine the outflow
velocity of the first and the last orifice.

3.11   Figure 3.33 is a blast pipe with four air outlets a, b, c, d. The section area of
the blast pipe is $50 \, \text{cm} \times 50 \, \text{cm}$. The section area of each air outlet is

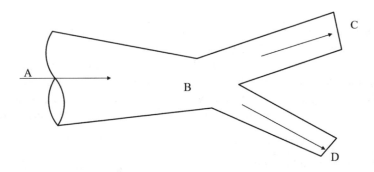

Fig. 3.30 Problem 3.8

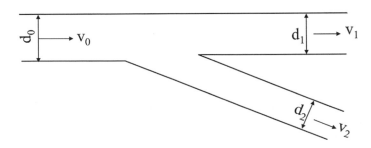

Fig. 3.31 Problem 3.9

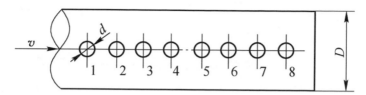

Fig. 3.32 Problem 3.10

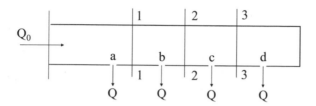

Fig. 3.33 Problem 3.11

Fig. 3.34 Problem 3.12

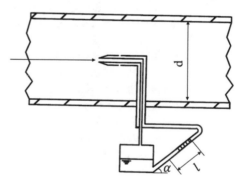

40 cm × 40 cm, and the air average velocity at the outlet is 5 m/s, try to determine the velocity and flow rate on the cross section 1-1, 2-2, and 3-3, respectively.

3.12 Figure 3.34 is a Pitot tube used for measuring the velocity u_{max} on the axis of the tube. An alcohol manometer is connected to the Pitot tube and $u_{max} = 1.2v$. The dimensions are $d = 200$ mm, $\sin \alpha = 0.2$ and $l = 75$ mm. The gas density and the alcohol density are 1.66 and 800 kg/m^3 respectively. Try to determine the mass flow rate of the gas.

3.13 A Pitot tube connected to a water manometer is used to measure the air velocity in the pipe, as shown in Fig. 3.35. The reading of the manometer is $h = 150$ mm H$_2$O. The air density is $\rho_a = 1.20$ kg/m^3, and the water density is $\rho - 1000$ kg/m^3. The velocity coefficient of Pitot tube is $c = 1$. Neglecting the energy loss, try to determine the air velocity u_0.

Fig. 3.35 Problem 3.13

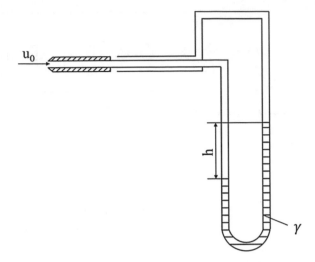

3.14 As shown in Fig. 3.36, Oil flows out from the vertical pipe. The pipe diameter
      is $d_1 = 10$ cm, and the velocity at the outlet is $v = 1.4$ m/s. Try to determine
      the velocity and the diameter of oil column which is $H = 1.5$ m below the
      outlet.

3.15 Figure 3.37 is an expanding water supply pipe. The dimensions are
      $d = 15$ cm, $D = 30$ cm. $p_A = 68.6$ kN/m², $p_B = 58.8$ kN/m², $h = 1$ m and
      $v_B = 0.5$ m/s. Try to determine the velocity v_A of point A, the head loss of
      section AB and the flow direction. $(\alpha = 1)$

3.16 As shown in Fig. 3.38, there is a pipe with variable diameter, and the angle
      between it and the horizontal plane is 45°. The diameter of cross section 1-1 is
      $d_1 = 200$ mm, and the diameter of 2-2 is $d_2 = 100$ mm. The distance between

Fig. 3.36 Problem 3.14

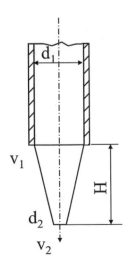

Fig. 3.37 Problem 3.15

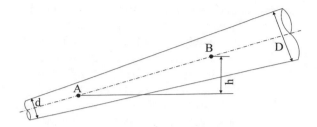

Fig. 3.38 Problem 3.16

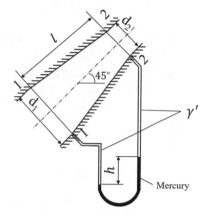

the two cross sections is $l = 2$ m. Oil flows through the pipe, and its specific weight γ' is 8820 N/m^3. The oil velocity of the section 1-1 is $v_1 = 2$ m/s, and the height difference in the mercury manometer is $h = 20$ cm. Try to determine: (1) the energy loss h_l between cross sections 1-1 and 2-2; (2) flow direction; (3) the pressure drop between sections 1-1 and 2-2.

3.17 As shown in Fig. 3.39, water flows from bottom to top. The dimensions are $d_1 = 300$ mm and $d_2 = 150$ mm. The dimensions of the manometer are $a = 80$ cm and $b = 10$ cm. Try to determine the flow rate.

Fig. 3.39 Problem 3.17

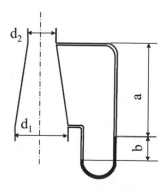

3.18 Figure 3.40 shows a centrifugal fan. The diameter of the suction pipe is $D = 200\,mm$, and there is a pressure measuring device with water connected to the suction pipe wall. The height is $h = 0.25$ m, and the specific weight of air is $\gamma_a = 12.64\,N/m^3$. Try to determine the flow rate of the air.

3.19 As shown in Fig. 3.41, there is a siphon with diameter $D = 20$ mm. Other dimensions are $H_1 = 2\,m$ and $H_2 = 6\,m$. Neglecting energy loss, how much pressure of point S is required to make the pipe start sucking water? And find the velocity and flow rate in the siphon then.

3.20 Figure 3.42 shows a Pitot tube connected to a horizontal pipeline. The dimensions are $D = 50\,mm$, $d = 25\,mm$. $p'_1 = 7.84\,kN/m^2$ and $Q = 2.7\,L/s$. Try to determine the height h_v of mercury column. (Neglecting loss)

3.21 A Pitot tube is installed to measure the flow rate of petroleum in the pipeline, as shown in Fig. 3.43. The diameter of the pipe is $d_1 = 20$ cm, and the diameter of the choke is $d_2 = 10$ cm. The oil density is $\rho = 850\,kg/m^3$, and the flow rate coefficient of Venturi tube is $\mu = 0.98$. The reading of the mercury manometer is $h = 15$ cm. Try to determine the flow rate Q of petroleum in the pipeline.

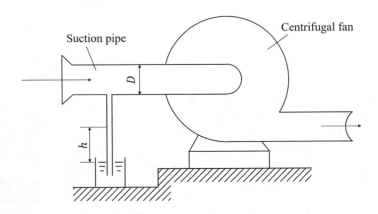

Fig. 3.40 Problem 3.18

Fig. 3.41 Problem 3.19

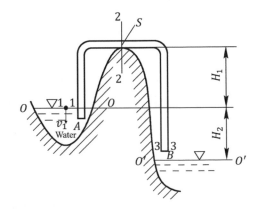

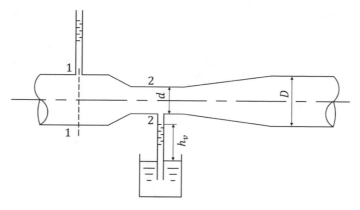

Fig. 3.42 Problem 3.20

Fig. 3.43 Problem 3.21

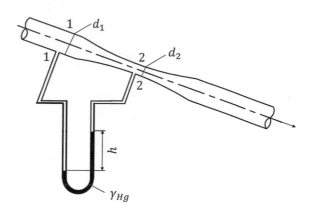

3.22 As shown in Fig. 3.44, a sealed water tank is used to supply water upward. The height is $h = 2$ m and the required flow rate is $Q = 15$ L/s. The diameter of the pipe is $d = 5$ cm, and the head loss is 50 cm H_2O. Try to determine the pressure required for the tank.

3.23 As shown in Fig. 3.45, air flows in from inlet a (its elevation is zero), and after burning, the waste gas flows through b, c (its elevation is 5 m), d (its elevation is 50 m) back to the atmosphere. The specific weight of the air is $\gamma_a = 11.8$ N/m³, and the specific weight of the waste gas is $\gamma = 5.9$ N/m³. The pressure drop from a to c is $9\gamma\frac{v^2}{2g}$, and the pressure drop from c to d is $20\gamma\frac{v^2}{2g}$. Try to determine the velocity v of waste gas at the outlet and the pressure p_c in point c.

3.24 As shown in Fig. 3.46, the diameter of the nozzle is $d = 75$ mm. The diameter of the hydraulic giant is $D - 150$ mm, of which the inclined angle is $\theta - 30°$. The reading of the pressure gauge is $h = 3$ m H_2O. Try to determine the

Fig. 3.44 Problem 3.22

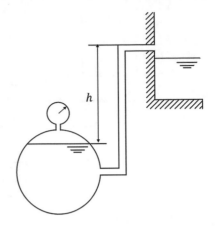

Fig. 3.45 Problem 3.23

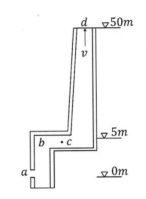

Fig. 3.46 Problem 3.24

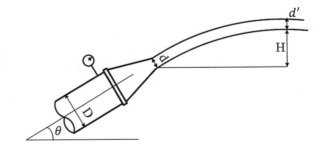

velocity v at the hydraulic giant outlet, maximum range H and the jet diameter d' at the highest point.

3.25 As shown in Fig. 3.47, there is a piping system with a pump. The flow rate is $Q = 1000\,\text{m}^3/\text{h}$, and the pipe diameter is $d = 150\,\text{mm}$. The total head loss of the pipeline is $h_{l1-2} = 25.4\,\text{m}\,H_2O$. The pump efficiency is $\eta = 80\%$, and the height difference between two reservoirs' water level is $h = 102\,\text{m}$. Try to determine head rise H and input power N of the pump.

Fig. 3.47   Problem 3.25

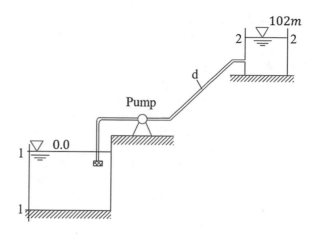

3.26   As shown in Fig. 3.48, there is a 45° elbow on the horizontal level, and the diameters of the inlet and the outlet are $d_1 = 600$ mm and $d_2 = 300$ mm, respectively. The relative pressure at the inlet is $p_1 = 40$ kPa, and the flow rate is $Q = 0.425 \, \text{m}^3/\text{s}$. Neglecting friction. Try to determine the force acting on the elbow by water.

3.27   As shown in Fig. 3.49, there is a pipe with diameter 150 mm, and two nozzles are connected to its end. The diameters of the two nozzles are 75 and 100 mm, respectively. The velocities at the two outlets are both 12 m/s. The pipe and nozzles are all on a same horizontal level. Neglecting friction, try to determine the magnitude and direction of the forces acting on the two nozzles by the water.

Fig. 3.48   Problem 3.26

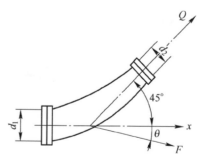

Fig. 3.49   Problem 3.27

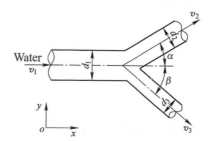

3.28   As shown in Fig. 3.50, the diameter of the vertical jet is $d = 7.5\,cm$, and the
        velocity is $v_0 = 12.2\,m/s$. The jet hits a dish whose gravity is 171.5 N. When
        the dish is balanced, try to determine y.

3.29   As shown in Fig. 3.51, the jet diameter is $d = 4\,cm$, and its velocity is
        $v = 20\,m/s$. The angel between the jet and the plane is $\theta = 30°$, and the
        velocity of the plane in its normal direction is $u = 8\,m/s$. Try to determine the
        force F acting on the plane in the normal direction.

3.30   As shown in Fig. 3.52, a board with a shape edge is inserted into the water jet,
        and the board is vertical to the jet. The velocity and flow rate of the jet is
        $v = 30\,m/s$ and $Q = 36$ L/s. The flow rate of the two distributaries are
        $Q_1 = \frac{1}{3}Q$ and $Q_2 = \frac{2}{3}Q$. Try to determine the deflection angle α of the jet and
        the force R exerting on the board by the jet.

Fig. 3.50   Problem 3.28

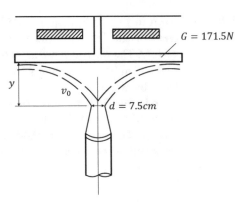

Fig. 3.51   Problem 3.29

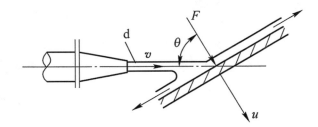

Fig. 3.52   Problem 3.30

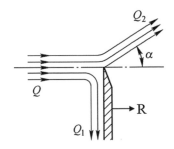

Fig. 3.53 Problem 3.31

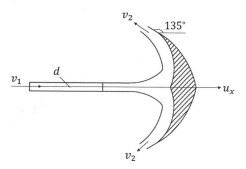

Fig. 3.54 Problem 3.32

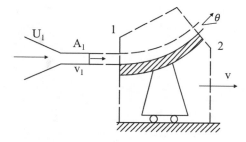

3.31 As shown in Fig. 3.53, a jet is lashing a blade. Assuming that $d = 10$ cm, $v_1 = v_2 = 20$ m/s and $\alpha = 135°$. Try to determine the force exerting on the blade when (1) the velocity of the blade $u_x = 0$; (2) $u_x = 10$ m/s.

3.32 As shown in Fig. 3.54, a jet is lashing a smooth blade and makes it moves along the horizontal direction at a constant velocity v. The deflection angle of the blade is θ. The section area at outlet of the jet is A_1, and the jet velocity is v_1. Considering the jet is steady compared with the blade, and neglecting gravity and friction, try to determine the force acting on the blade by the jet.

References

1. Richardson, S.M.: Fluid Mechanics. Hemisphere Publishing Corporation, New York (1989)
2. Aris, R.: Vectors, Tensors, and the Basic Equations of Fluid Mechanics. Dover Publication, New York (1962)
3. Spurk, J.: Fluid Mechanics. Springer, Heidelberg (1997)
4. Xie, Z.: Engineering Fluid Mechanics, 4th edn. Metallurgical Industry Press, Beijing (2014)
5. Song, H.: Engineering Fluid Mechanics and Environmental Application. Metallurgical Industry Press, Beijing (2016)
6. Ferziger, H.J.: Peric, Computational Methods for Fluid Dynamics, 3rd edn. Springer, Heidelberg

Head Loss of Incompressible Viscous Flow

4

Abstract

Actual fluids have different flow regimes due to viscosity. The regimes include laminar flow and turbulent flow. The regimes, viscosity, and pipe wall surface have influence on flow resistance, which leads to head loss for fluids flow. In this chapter, we will firstly introduce types of head loss (friction loss and minor head loss) and two regimes of fluids flow. Then we will study characteristics of laminar flow and turbulent flow in circular pipe. Furthermore, we will present the definition of friction factor and discuss ways to determine value of friction factor in order to calculate friction loss. Finally, we will give determination of minor head loss.

Keywords

Friction loss · Reynolds number · Laminar flow · Turbulent flow
Nikuradse tests · Moody chart · Boundary layer

4.1 Types of Head Loss

4.1.1 The Hydraulic Diameter

There are two factors affecting the magnitude of flow resistance: cross-section area A and wetted perimeter χ, which is the length of wall in contact with the flowing fluid on any cross section.

Hydraulic diameter D or hydraulic radius R is often used to reflect the comprehensive influence on flow resistance with two factors above, defined as $D = \frac{4A}{\chi}, R = \frac{A}{\chi}$ separately. The hydraulic diameter will equal the duct diameter for a circular cross section. For a circular duct with diameter d and radius r, $A = \pi r^2$ and $\chi = 2\pi r$, then we have

$$R = \frac{A}{\chi} = \frac{\pi r^2}{2\pi r} = \frac{r}{2} = \frac{d}{4} \quad \text{and} \quad D = \frac{4A}{\chi} = 2r = d$$

4.1.2   Friction Loss and Minor Head Loss

If the streamlines are straight and parallel to each other, then this kind of flow is called uniform flow, otherwise, it is called nonuniform flow.

In fluid flow, friction loss (or skin friction) is the loss of pressure or "head" that occurs in pipe or duct flow due to the effect of the fluid viscosity. The term refers to the power lost in overcoming the friction between two moving surfaces. The characteristic of friction loss is that it is proportional to the pipe length. Friction loss per unit weight of flowing fluid in pipe is denoted by h_f.

The flow in a piping system may be required to pass through a variety of fittings, elbow, or abrupt changes in area. Additional head losses are encountered, primarily as a result of flow separation. Energy eventually is dissipated by violent mixing in the separated zones. Therefore, these losses will be minor loss which is denoted by h_r.

For the entire piping system, its total head loss h_l should be the summation of friction losses and minor losses, that is

$$h_l = \sum h_f + \sum h_r \tag{4.1}$$

4.2   Two Regimes of Viscous Flow

4.2.1   Reynolds Experiment

Viscous flows generically fall into two categories, laminar and turbulent, but the boundary between them is imperfectly defined. The basic difference between the two categories is phenomenological and was dramatically demonstrated in 1883 by Reynolds [1–3], who injected a thin stream of dye into the flow of water through a tube (Fig. 4.1a).

Reynolds's experiment was to distinguish between laminar and turbulent flows. At low flow rate [the upper drawing (Fig. 4.1b)], the pipe flow was laminar and the dye filament moved smoothly through the pipe. At high flow rate [the lower drawing (Fig. 4.1c)], the flow became turbulent and the dye filament was mixed throughout the cross section of the pipe. At low flow rate, the dye stream was observed to follow a well-defined straight path, indicating that the fluid moved in parallel layers (laminar) with no unsteady macroscopic mixing or overturning motion of the layers. Such smooth orderly flow is called laminar flow. However, if the flow rate was increased beyond a certain critical value, the dye streak broke up into irregular filaments and spread throughout the cross section of the tube,

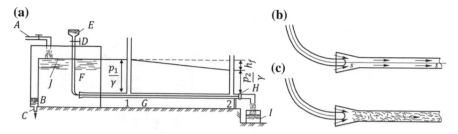

Fig. 4.1 Reynolds Experiment [4] A: water supply pipe; B: the water tank; C: drainage pipe; D: a valve to control dye velocity; E: dye; F, G: glass pipe; H: a valve to control flow velocity; I: measuring tank; J: overflow plate

indicating the presence of unsteady, apparently chaotic three-dimensional macroscopic mixing motions. Such irregular disorderly flow is called a turbulent flow.

If the experiment is conducted in a reverse process, it can be found that the transformation of flow regime is: turbulent flow to laminar flow. When without external disturbance, the flow velocity at which laminar flow transforms to transient state is called the upper critical velocity, denoted by v'_c; and that from turbulent flow to transient state is called the lower critical velocity, denoted by v_c shown in Fig. 4.2. The upper critical velocity is higher than the lower critical velocity. In most cases, the lower critical velocity is used as the criterion for judging flow pattern, and is referred to as the critical velocity.

According to the experiment, we can draw a conclusion

(1) When $v > v'_c$, fluid flows in turbulent flow;
(2) When $v < v_c$, fluid flows in laminar flow;
(3) When $v_c < v < v'_c$, the flow regime is unstable, and the original laminar or turbulent flow may be maintained.

4.2.2 Criteria for Flow Regime

Based on a great many experiments, Reynolds proposed Reynolds number as the criterion to judge flow regime. For circular pipes, Reynolds number is defined as

$$Re = \frac{vd}{v},\tag{4.2}$$

where v is the fluid velocity, d is the inner diameter of the pipe and v is kinetic or kinematic viscosity of fluid.

Reynolds numbers corresponding to the upper and lower critical velocity are expressed as the upper and lower critical Reynolds number respectively

$$Re'_c = \frac{\rho v'_c d}{\mu} = \frac{v'_c d}{v}, \ Re_c = \frac{\rho v_c d}{\mu} = \frac{v_c d}{v}$$

Fig. 4.2 Critical velocity in
Reynolds experiment [5]

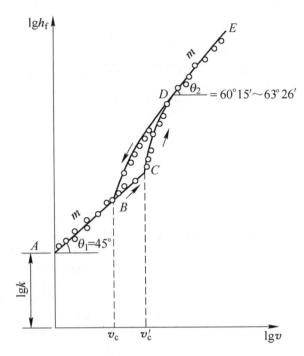

Thus, we may utilize upper and lower Reynolds number to determine flow pattern, namely,

(1) When $Re < Re_c$, laminar flow;
(2) When $Re_c < Re < Re_c'$, transient state;
(3) When $Re > Re_c'$, turbulent flow.

Many experiments demonstrate that the lower critical Reynolds number Re_c is always equal to 2320 for circular duct. Conventionally, the lower critical Reynolds number Re_c is used as the criterion between laminar and turbulent flow, that is

(1) When $Re \leq 2320$, laminar flow;
(2) When $Re > 2320$, turbulent flow.

In practical engineering, the external disturbance is very easy to make the fluid in turbulent, so the practical critical Reynolds number will be smaller than 2320 with value following:

$$Re_c = 2000$$

When the fluid flows in a noncircular pipe, the hydraulic radius can be used as the characteristic length, and the critical Reynolds number is

$$Re_c = 500$$

For open channel flow, it is easier to reach turbulent regime due to external influence. Frequently the critical Reynolds number is used in engineering calculation as follows:

$$Re_c = 300$$

Example 4.1 The water with temperature $t = 15\ °C$ flows in a pipe with diameter $d = 100$ mm, and the flow rate is $Q = 15$ L/s. There is another rectangular open channel with width 2 m, and water depth 1 m, and the average velocity is 0.7 m/s. The water temperature is same as above. Try to determine flow patterns of the two cases.

Solution

When water temperature is 15 °C, we can know that the kinematic viscosity of water is $v = 1.141 \times 10^{-6}\ m^2/s$ according to Table 1.2.

(1) The flow velocity of water in circular pipe is

$$v = \frac{Q}{A} = \frac{15 \times 10^{-3}}{\frac{\pi \times 0.1^2}{4}} = 1.911\ \text{m/s}$$

The Reynolds number of the flow in circular pipe is

$$Re = \frac{vd}{v} = \frac{1.911 \times 0.1}{1.141 \times 10^{-6}} = 167632 \gg 2000$$

Therefore, the water flows in turbulent flow

(2) The hydraulic radius of the open channel is

$$R = \frac{A}{\chi} = \frac{2 \times 1}{2 + 2 \times 1} = 0.5\ \text{m}$$

$$Re = \frac{vR}{v} = \frac{0.7 \times 0.5}{0.0114 \times 10^{-4}} = 30701 \gg 300$$

Therefore, the water flows in turbulent flow

Example 4.2 The water with temperature $t = 15\ °C$ and kinematic viscosity $v = 0.0114\ cm^2/s$, flows in a pipe with diameter $d = 20$ mm, and the velocity is $v = 8$ cm/s. Try to determine flow pattern of the water. What can you do to change flow pattern?

Solution

The Reynolds number of water flow in the pipe is

$$Re = \frac{vd}{v} = \frac{8 \times 2}{0.0114} = 1403.5 < 2000$$

So the water flows in laminar flow. To change the flow pattern, we can

(1) Increase velocity

If $Re_c = 2000$ is adopted and the viscosity of water is unchanged, the velocity of water should be

$$v = \frac{Re_c v}{d} = \frac{2000 \times 0.0114}{2} = 11.4 \text{ cm/s}$$

Therefore, the flow pattern of water will become turbulent when the flow velocity increases to 11.4 cm/s.

(2) Increase water temperature to reduce the viscosity of water

If $Re_c = 2000$ is adopted, and the velocity of water is unchanged, the kinematic viscosity of water is

$$v = \frac{vd}{Re_c} = \frac{8 \times 2}{2000} = 0.008 \text{ cm}^2/\text{s}$$

According to Table 1.2: when water temperature is $t = 30$ °C, $v = 0.00804$ cm^2/s; when water temperature is $t = 35$ °C, $v = 0.00727$ cm^2/s.

Therefore, if we can increase the water temperature to 31 °C, the flow pattern of the water can be turbulent.

4.3 Laminar Flow in Circular Pipe

4.3.1 Two Methods for Laminar Flow Analysis

4.3.1.1 N-S Equation Analysis Method

Steady laminar flow of incompressible fluid in the circular pipe has five characteristics as follows:

(1) Only one-dimensional flow in y direction. Establish the coordinate system as shown in Fig. 4.3. The y axis should coincide with central line of circular pipe. Since there is only one-dimensional flow in y direction, $u_y \neq 0, u_x = u_z = 0$. The N-S equation can be rewritten as

Fig. 4.3 Control volume for
analysis of laminar flow

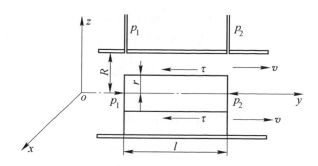

$$
\left.
\begin{aligned}
X - \frac{1}{\rho}\frac{\partial p}{\partial x} &= 0 \\
Y - \frac{1}{\rho}\frac{\partial p}{\partial y} + v\left(\frac{\partial^2 u_y}{\partial x^2} + \frac{\partial^2 u_y}{\partial y^2} + \frac{\partial^2 u_y}{\partial z^2}\right) &= \frac{\partial u_y}{\partial y}u_y + \frac{\partial u_y}{\partial t} \\
Z - \frac{1}{\rho}\frac{\partial p}{\partial z} &= 0
\end{aligned}
\right\}
\qquad (4.3)
$$

(2) Steady and incompressible fluid flow. For steady flow, $\frac{\partial u_y}{\partial t} = 0$. In terms of the
continuity equation of incompressible fluid $\frac{\partial u_x}{\partial x} + \frac{\partial u_y}{\partial y} + \frac{\partial u_z}{\partial z} = 0$, it can be
obtained that $\frac{\partial u_y}{\partial y} = 0$ under condition (4.1), so $\frac{\partial^2 u_y}{\partial y^2} = 0$.

(3) Rotational symmetry of velocity distribution. On cross section of the circular
duct, the flow velocity at each point is different but rotational symmetrical, so
the velocity on the x axis and z axis can be indicated in polar coordinates with r.
We have $\frac{\partial^2 u_y}{\partial x^2} = \frac{\partial^2 u_y}{\partial z^2} = \frac{\partial^2 u_y}{\partial r^2} = \frac{\mathrm{d}^2 u_y}{\mathrm{d}r^2}$.

(4) Linear pressure drop along flow direction in pipe with constant diameter. Due
to friction between fluid layers or flow layer and pipe boundary, the pressure is
gradually decreasing along the flow direction. In pipeline of constant radius this
decline is linear, and the change rate of pressure per unit length $\frac{\partial p}{\partial y}$ can be
expressed as $\frac{\partial p}{\partial y} = \frac{\mathrm{d}p}{\mathrm{d}y} = -\frac{p_1 - p_2}{l} = -\frac{\Delta p}{l}$, in which the negative sign indicates that
the pressure drops along the flow direction.

(5) No mass force in y direction for practical pipe. For practical horizontal pipes,
the mass force in x, y, z direction can be shown as $X = Y = 0$ and $Z = -g$.

According to above five characteristics, Eq. (4.3) can be simplified as

$$
\frac{\Delta p}{\rho l} + 2v\frac{\mathrm{d}^2 u_y}{\mathrm{d}r^2} = 0
\qquad (4.4)
$$

Integrating Eq. (4.4), we have

$$
\frac{\mathrm{d}u_y}{\mathrm{d}r} = -\frac{\Delta p}{2\mu l}r + C
\qquad (4.5)
$$

There exists the maximum velocity in the center of the circular pipe $r = 0$, so $\frac{du_y}{dr} = 0$. and then the integral constant C equals 0, we have

$$\frac{du_y}{dr} = -\frac{\Delta p}{2\mu l} r \qquad (4.6)$$

This is the ordinary differential equation of laminar flow in circular pipe.

4.3.1.2   Total Force Analysis Method

As shown in Fig. 4.3, select a cylindrical control volume with length l and radius r, and it is in steady flow state.

The forces acting on the control volume include the pressure on the base area $(p_1 - p_2)\pi r^2$ and the friction on the lateral area or side area of cylinder $\tau 2\pi r l$. The projection of the external force acting on the control volume along y direction equals zero $\sum F_y = 0$, therefore it can be obtained that $(p_1 - p_2)\pi r^2 - \tau 2\pi r l = 0$.

In terms of the Newton's viscosity law $\tau = -\mu \frac{du_y}{dr}$, it can be obtained that $\frac{du_y}{dr} = -\frac{p_1 - p_2}{2\mu l} r = -\frac{\Delta p}{2\mu l} r$.

The results of the two methods are same. Compared with the first method, the second one is much easier to understand including clear physical description. The two methods are available only with the conditions of one-dimensional, steady, rotational symmetry and uniform flow.

4.3.2   Velocity Profile and Shear Stress Distribution of Laminar Flow in Pipes

Integrating on both sides of Eq. (4.6), we have

$$u_y = -\frac{\Delta p}{4\mu l} r^2 + C$$

According to boundary conditions, when $r = R, u_y = 0$, so $C = \frac{\Delta p}{4\mu l} R^2$. Therefore, the velocity profile of laminar flow in pipe can be expressed as

$$u_y = \frac{\Delta p}{4\mu l} (R^2 - r^2) \qquad (4.7)$$

The above equation indicates that the velocity distribution for laminar flow in a pipe is parabolic across the section with the maximum velocity at the center of the pipes. The parabolic profile for fully developed laminar pipe flow is sketched in Fig. 4.4.

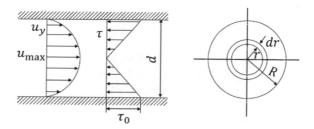

Fig. 4.4 Velocity profile and shear stress distribution of laminar flow in pipes

At $r = 0$,

$$u_{max} = \frac{\Delta p}{4\mu l} R^2 \tag{4.8}$$

According to the Newton's viscosity law (stress equation)

$$\tau = \pm \mu \frac{du_y}{dr} = -\mu \frac{du_y}{dr} = \frac{\Delta p r}{2l} \tag{4.9}$$

The above equation shows that shear stress on cross section in laminar flow is proportional to the pipe radius, as shown in Fig. 4.4, which is K-shaped for shear stress distribution.

When $r = R$, the maximum shear stress is obtained on pipe wall

$$\tau_0 == \frac{\Delta p R}{2l} \tag{4.10}$$

4.3.3 Flow Rate and Average Velocity of Laminar Flow in Pipes

Selecting an arbitrary point at r in a pipe with infinitesimal width dr, we have cross section area $dA = 2\pi r dr$. Then we can get the pipe flow rate as follows:

$$Q = \int_A u_y dA = \int_0^R \frac{\Delta p}{4\mu l} (R^2 - r^2) 2\pi r dr = \frac{\pi \Delta p R^4}{8\mu l} = \frac{\pi \Delta p d^4}{128\mu l} \tag{4.11}$$

The above equation is called Hagen–Poiseuille equation. The calculation results in terms of Hagen–Poiseuille equation have great agreement with precise experimental tests, which corresponds to the practical result perfectly.

Hagen–Poiseuille equation is also utilized to measure fluid viscosity according to the equation as follows:

$$\mu = \frac{\pi \Delta p d^4}{128 l Q} = \frac{\pi \Delta p d^4 t}{128 l V}$$

The fluid viscosity can be obtained with pipe inner diameter d and length l, the measured pressure drop Δp and flow volume V within time t.

The average velocity of the circular pipe is

$$v = \frac{Q}{A} = \frac{\pi \Delta p R^4}{8 \mu l \cdot \pi R^2} = \frac{\Delta p}{8 \mu l} R^2 \qquad (4.12)$$

It can be known from Eqs. (4.8) and (4.12) that $u_{max} = 2v$. It means that the maximum velocity of laminar flow in circular pipe is twice average velocity, so the velocity distribution is quite nonuniform.

4.3.4 Friction Head Loss in Laminar Flow

According to Bernoulli's equation, we have

$$h_f = \left(z_1 + \frac{p_1}{\gamma} + \frac{v_1^2}{2g} \right) - \left(z_2 + \frac{p_2}{\gamma} + \frac{v_2^2}{2g} \right)$$

The pipe is horizontal $z_1 = z_2$ with the average velocity $v_1 = v_2$. Combining with Eq. (4.12), the friction head loss can be shown as

$$h_f = \frac{p_1 - p_2}{\gamma} = \frac{\Delta p}{\gamma} = \frac{8 \mu l v}{\gamma R^2} = \frac{32 \mu l v}{\gamma d^2} \qquad (4.13)$$

The above equation can be changed for engineering calculation into following:

$$h_f = \frac{32 \mu l}{\gamma d^2} v = \frac{64}{\frac{\rho v d}{\mu}} \frac{l}{d} \frac{v^2}{2g} = \frac{64}{Re} \frac{l}{d} \frac{v^2}{2g} = \lambda \frac{l}{d} \frac{v^2}{2g} \qquad (4.14)$$

where $\lambda = \frac{64}{Re}$ is called the friction factor in laminar flow, which is only related to Reynolds number Re. Equation (4.14) is commonly used to calculate the friction loss, which is also called Darcy–Weisbach equation.

When fluid is transported by the pump in pipe, it is necessary to obtain the power consumed due to frictional drag. If the fluid specific weight γ and flow rate Q are known, then the consumed power of laminar flow in circular pipe with length l is

$$N = \gamma Q h_f = \gamma Q \frac{\lambda l}{d} \frac{v^2}{2g} \qquad (4.15)$$

4.3.5   The Entrance Region

Figure 4.5 illustrates laminar flow in the entrance region of a circular pipe. The flow has uniform velocity U_0 at the pipe entrance. Because of the no-slip condition on the wall, we know that the velocity on the wall must be zero along the entire length of the pipe. A boundary layer develops along the walls of the channel. The solid surface exerts a retarding shear force on the flow; thus, the speed of the fluid in the neighborhood of the surface is reduced. At successive sections along the pipe in this entry region, the effect of the solid surface is felt farther out into the flow.

For incompressible flow, mass conservation requires that, as the speed close to the wall is reduced, the speed in the central frictionless region of the pipe must increase slightly to compensate; for this inviscid central region, then, the pressure (as indicated by the Bernoulli equation) must also drop somewhat.

Sufficiently far from the pipe entrance, the boundary layer developing on the pipe wall reaches the pipe centerline and the flow becomes entirely viscous. The velocity profile shape then changes slightly after the inviscid core disappears. When the profile shape no longer changes with increasing distance x, the flow is called fully developed. The distance downstream from the entrance to the location at which fully developed flow begins is called the entrance length. The actual shape of the fully developed velocity profile depends on whether the flow is laminar or turbulent. In Fig. 4.5 the profile is shown qualitatively for a laminar flow. Although the velocity profiles for some fully developed laminar flows can be obtained by simplifying the complete equations of motion, turbulent flows cannot be so treated.

For laminar flow, it turns out that entrance length, L, is a function of Reynolds number,

$$\frac{L}{D} \cong 0.06 \frac{\rho v D}{\mu},$$

where $v = \frac{Q}{A}$ is the average velocity (because flow rate $Q = Av = AU_0$, we have $v = U_0$).

Laminar flow in a pipe may be expected only for Reynolds numbers less than 2320. Thus, the entrance length for laminar pipe flow may be as long as

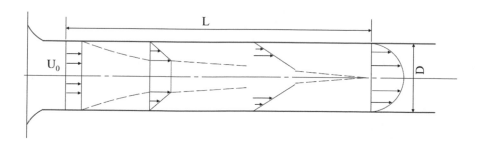

Fig. 4.5 Flow in the entrance region of a pipe

$$L \cong 0.06 R_e D \le (0.06)(2320)D = 139D$$

or nearly 140 pipe diameters.

Example 4.3 A heavy oil with specific weight 9.31 kN/m³ is transported in a pipeline with length $L = 1000$ m and diameter $d = 300$ mm, and the flow rate is $G = 2300$ kN/h. Try to determine head loss if the oil temperature is 10 °C ($v = 25$ cm²/s) and 40 °C ($v = 1.5$ cm²/s) respectively.

Solution

The volumetric flow rate of the heavy oil is

$$Q = \frac{G}{\gamma} = \frac{2300}{9.31 \times 3600} = 0.0686 \text{ m}^3/\text{s}$$

The average velocity of heavy oil is

$$v = \frac{Q}{A} = \frac{0.0686}{\frac{\pi}{4} \times 0.3^2} = 0.971 \text{ m/s}$$

The Reynolds number at 10 °C is

$$Re_1 = \frac{vd}{v} = \frac{0.971 \times 0.3}{25 \times 10^{-4}} = 116.5 < 2000$$

The Reynolds number at 40 °C is

$$Re_2 = \frac{vd}{v} = \frac{0.971 \times 0.3}{1.5 \times 10^{-4}} = 1942 < 2000$$

The flow patterns of heavy oil in two cases are laminar, and the corresponding friction losses can be obtained by Darcy Eq. (4.14).

$$h_{f1} = \frac{\lambda_1 l \, v^2}{d \, 2g} = \frac{64}{Re_1} \frac{l \, v^2}{d \, 2g} = \frac{64}{116.5} \times \frac{1000}{0.3} \times \frac{0.971^2}{2 \times 9.8} = 88.1 \text{ m}$$

$$h_{f2} = \frac{\lambda_2 l \, v^2}{d \, 2g} = \frac{64}{Re_2} \frac{l \, v^2}{d \, 2g} = \frac{64}{1942} \times \frac{1000}{0.3} \times \frac{0.971^2}{2 \times 9.8} = 5.28 \text{ m}$$

It can be seen from the calculation that the head loss of heavy oil flow at 40 °C is less than that at 10 °C.

4.4 Turbulent Flow in Circular Pipe

4.4.1 Parameters Description in Turbulent Flow

As shown in Fig. 4.6, in laminar flow, the fluid particles pass through point m(or point n) will follow a certain path to point m'(or point n'). However, in turbulent flow, the fluid particles pass through point m will follow tortuous and rambling paths to point C (or point n'). In turbulent flow kinetic parameters u and p of each particle in flow field are irregular with time, which is referred to as fluctuation phenomenon. This fluctuation phenomenon of flow variables with time indicates that turbulent flow is not steady flow, which leads to challenges in turbulent flow research.

However, this kind of flow still follows a certain rule for a long time. All physical parameters can be dealt with by mean values and fluctuation.

Mean velocity $\overline{u_x}$ is the average of instant velocity u_x during time interval T. The integral area between the curve of instant velocity u_x and the t axis to time T should be equal to the product of mean velocity $\overline{u_x}$ and interval T, as shown in Fig. 4.7, namely

$$\int_0^T u_x \mathrm{d}t = T\overline{u_x}$$

So

$$\overline{u_x} = \frac{1}{T}\int_0^T u_x \mathrm{d}t \tag{4.16}$$

Fig. 4.6 Turbulent flow

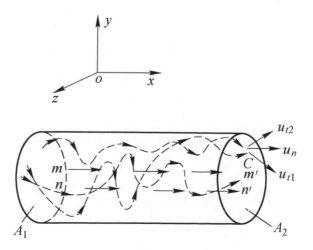

Fig. 4.7 Sketch of mean
velocity and fluctuation

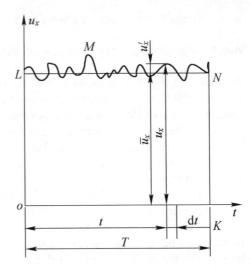

In turbulent flow the mean velocity is reliable in terms of long enough time. If the time interval T is too short, there is no relationship between mean velocity and time interval T.

It can be seen from Fig. 4.7 that the instant velocity u_x can be shown as the sum of mean velocity $\overline{u_x}$ and fluctuation u'_x

$$u_x = \overline{u_x} + u'_x \tag{4.17}$$

Substituting Eq. (4.17) into (4.16), it yields

$$\overline{u_x} = \frac{1}{T}\int_0^T u_x dt = \frac{1}{T}\int_0^T (\overline{u_x} + u'_x)dt = \frac{1}{T}\int_0^T \overline{u_x}dt + \frac{1}{T}\int_0^T u'_x dt = \overline{u_x} + \overline{u'_x} \tag{4.18}$$

So

$$\overline{u'_x} = \frac{1}{T}\int_0^T u'_x dt = 0 \tag{4.19}$$

This indicates that the mean magnitude of velocity fluctuation equals zero. The above method can also be applied to other fluid parameters as follows:

$$\begin{cases} u_y = \overline{u_y} + u'_y \\ u_z = \overline{u_z} + u'_z \\ p = \overline{p} + p' \end{cases} \tag{4.20}$$

Obviously, for one-dimensional flow, $\overline{u_y}$ and $\overline{u_z}$ should be zero, and u_y, u_z should be equal to u'_y and u'_z, respectively.

4.4.2 Mixing Length Theory

The mixing length theory of turbulence was proposed by Prandtl in 1925. It can reasonably explain the influence on average and fluctuation flow and is the basis for shear stress, velocity, and friction calculation in turbulent flow, so it is widely used in industry.

Shear stress in turbulent flow is composed of two parts: viscous shear stress caused by relative movement between neighboring layer, which is generated by mean velocity, and Reynolds shear stress caused by collision of fluid particles, which is generated by velocity fluctuation.

Therefore, the total shear stress in turbulent flow may be expressed as

$$\tau = \tau_1 + \tau_2, \tag{4.21}$$

where τ_1 is viscous shear stress, τ_2 is Reynolds shear stress or additional shear stress.

Viscous shear stress can be expressed according to Newton's viscosity law as

$$\tau_1 = \mu \frac{\mathrm{d}\overline{u}}{\mathrm{d}y}$$

Reynolds shear stress can be expressed directly in terms of mixing length theory as follows:

$$\tau_2 = \rho l^2 \left(\frac{\mathrm{d}\overline{u}}{\mathrm{d}y}\right)^2, \tag{4.22}$$

where l is called mixing length without physical meaning, $\overline{u}$ is the mean velocity.

So total shear stress in turbulent flow can be expressed as

$$\tau = \tau_1 + \tau_2 = \mu \frac{\mathrm{d}\overline{u}}{\mathrm{d}y} + \rho l^2 \left(\frac{\mathrm{d}\overline{u}}{\mathrm{d}y}\right)^2 \tag{4.23}$$

The magnitudes of τ_1 and τ_2 differ in different cases associated with Reynolds number. When Reynolds number is small enough, viscous shear stress τ_1 is dominated in the pipe flow. Conversely, Reynolds shear stress τ_2 would be dominated and τ_1 can be neglected with large enough Reynolds number.

4.4.3 Velocity Distribution of Turbulent Flow in Pipes

4.4.3.1 Velocity Distribution

According to Karman experiment, the relationship between mixing length l and the distance from fluid layer to pipe wall y can be approximately expressed as

$$l = ky\sqrt{1 - \frac{y}{R}}, \qquad (4.24)$$

where R is pipe radius. When $y \ll R$, namely, near the wall

$$l = ky, \qquad (4.25)$$

where k is Karman constant found to be 0.4. Therefore, formula (4.22) can be written as

$$\tau_2 = \rho k^2 y^2 \left(\frac{\mathrm{d}u}{\mathrm{d}y}\right)^2 \qquad (4.26)$$

For simplicity, the time-averaged symbols in the above equation are neglected, and we only discuss fully developed turbulent. It can be obtained from Eq. (4.26) that

$$\mathrm{d}u = \frac{1}{k}\sqrt{\frac{\tau}{\rho}\frac{\mathrm{d}y}{y}} \qquad (4.27)$$

Replace τ with friction resistance τ_0 on the wall, and assume that $\sqrt{\frac{\tau_0}{\rho}} = v_*$, which is called shear stress velocity, the above equation can be transformed into

$$\mathrm{d}u = \frac{v_*}{k}\frac{\mathrm{d}y}{y}$$

Integrate the above equation

$$u = \frac{v_*}{k}\ln y + C \qquad (4.28)$$

The above formula reflects velocity distribution of turbulent flow deduced by mixing length theory. It can be seen that in turbulent flow, the curve of velocity distribution on cross section is logarithm-shaped, as shown in Fig. 4.8. According to measurement tests, the average velocity v is 0.75–0.87 times comparing with the maximum velocity u_{max} at the center line.

The logarithmic distribution of turbulent velocity is accurate relatively, but the formula is too difficult to utilize. According to the experimental curve of turbulent flow in smooth pipes, the velocity distribution of turbulent flow can also be approximately expressed as a simple exponential formula:

Fig. 4.8 Velocity
distribution of turbulent flow

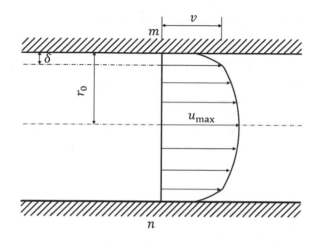

$$\frac{u_x}{u_{max}} = \left(\frac{y}{R}\right)^n \tag{4.29}$$

For different Re, the corresponding exponent n is different, ranging $n = 1/10 - 1/4$.

4.4.3.2 Laminar Sublayer, Hydraulically Smooth Pipe, and Hydraulically Rough Pipe

There is a very thin fluid layer with laminar flow in vicinity of pipe wall in turbulent flow, since the friction effect of pipe wall, the velocity is very low and inertia is very small, so that flow in this layer maintains laminar flow. This layer is defined as laminar sublayer.

Fluid particles in region beyond sublayer collide and mix with another showing the characteristic of turbulent flow, which is called turbulent core region. The thickness δ of laminar sublayer is not constant and depends on kinematic viscosity, fluid velocity, pipe diameter and friction factor. Its approximate calculation formula can be expressed as

$$\delta = \frac{32.8d}{Re\sqrt{\lambda}} \tag{4.30}$$

For crude oil with high viscosity, the thickness δ of laminar sublayer is only several millimeters. For common fluids, δ ranges from 10^{-2} to 10^{-1} mm. Due to existence of laminar sublayer, there are great influences on properties of pipe flow, such as thermal conductivity, flow resistance, etc.

It can be known from the previous statement that for laminar flow in pipes, friction factor is merely the function of Reynolds number. But in turbulent flow, besides Reynolds number, roughness of pipe wall is another important factor influencing friction factor. Roughness may be classified as surface roughness and relative roughness: surface roughness is the average thickness Δ of pipe wall roughness, and relative roughness is the ratio of surface roughness Δ to pipe diameter d, as shown in Fig. 4.9.

When $\delta > \Delta$, rough surface on pipes are all submerged inside laminar sublayer, it seems that fluid flows through relative smooth wall of pipe. Therefore, frictional loss has no relationship with roughness, in this case the pipe is called hydraulically smooth pipe, or smooth pipe for short.

When $\delta < \Delta$, rough surface on pipe wall protrudes into turbulent core region, and consumes couples of energy due to disturbance and eddies generation. In this case, frictional loss depends on pipe wall roughness, and the pipe is called hydraulically rough pipe, or rough pipe for short.

4.4.4 Head Loss of Turbulent Pipe Flow

Since uniform flow is discussed, friction resistance τ_0 on the wall can be calculated by formula (4.10), namely, $\tau_0 = \frac{\Delta p R}{2l} = \frac{\Delta p d}{4l}$, and $h_f = \frac{\Delta p}{\rho g}$, therefore

$$h_f = \frac{4\tau_0 l}{\rho g d} \tag{4.31}$$

τ_0 is too complicated to give an accurate expression for determination. We only can obtain the value of τ_0 by experimental and empirical approaches.

Couples of experimental results indicate that τ_0 is related to average velocity v, Reynolds number Re, and the ratio of pipe surface roughness Δ to radius r, so it can be expressed as

$$\tau_0 = f(Re, v, \Delta/r) = f_1(Re, \Delta/r)v = Fv^2 \tag{4.32}$$

Substituting the above equation into Eq. (4.31)

$$h_f = \frac{4Fv^2}{\rho g}\frac{l}{d} = \frac{8F}{\rho}\frac{l}{d}\frac{v^2}{2g} = \frac{\lambda l}{d}\frac{v^2}{2g} \tag{4.33}$$

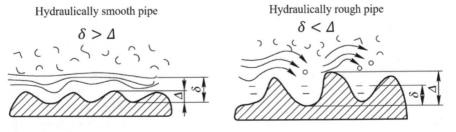

Fig. 4.9 Hydraulically smooth pipe and hydraulically rough pipe

where $\lambda = \frac{8F}{\rho} = f_1\left(Re, \frac{\Delta}{r}\right)$ is called friction factor of turbulent flow, which can only be determined by experiments.

4.5 Determination of Friction Factor in Circular Pipe

4.5.1 Calculation Equation

Because of the complexity of turbulent flow, the calculation equation for frictional loss in turbulent flow can only be set up by applying dimensional analysis. The calculation equation for frictional loss in turbulent flow is the same as that in laminar flow in form, namely

$$h_f = \frac{\lambda l}{d}\frac{v^2}{2g}$$

What is different is the magnitude of friction factor λ. For laminar flow

$$\lambda = \frac{64}{Re}$$

For turbulent flow, λ is the function of Reynolds number Re and relative roughness Δ/d, that is

$$\lambda = f\left(Re, \frac{\Delta}{d}\right)$$

4.5.2 Nikuradse Tests

German dynamicist and engineer Nikuradse conducted experimental determination on friction factor and cross-section velocity distribution in pipe flow in 1933. The experimental equipment is similar to Reynolds experiments.

Nikuradse simulated roughness by gluing uniform sand grains onto the inner walls of the pipes. He then measured the pressure drops and flow rates and correlated friction factor versus Reynolds number. The curve $\lambda = f\left(Re, \frac{\Delta}{d}\right)$ presented in such a way is called Nikuradse curve graph, as shown in Fig. 4.10.

According to the variation characteristic of λ, the chart of Nikuradse tests is divided into five regions.

Region I: Laminar flow region (line ab). When $Re < 2320$(lg $Re < 3.36$), λ is free from relative roughness Δ/d, and the magnitude of friction factor agrees with the equation $\lambda = \frac{64}{Re}$.

Region II: Transition region of flow regime (line bc). When $2320 < Re < 4000$ (lg $Re = 3.36 \sim 3.6$), λ only depends on Re and is independent from Δ/d.

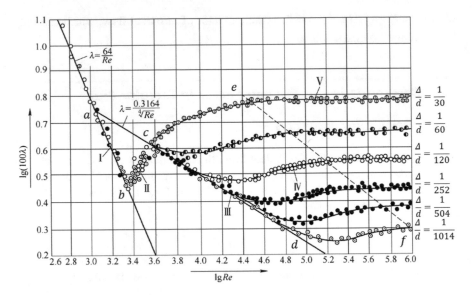

Fig. 4.10 Chart of Nikuradse tests [5]

Region III: Turbulent flow in smooth pipes (line cd). When $4000 < Re < 22.2\left(\frac{d}{\Delta}\right)^{\frac{8}{7}}$, flow is in turbulent regime where only Re influences on λ rather than Δ/d.

The formula for calculating λ in this region is as follows:

① When $4000 < Re < 10^5$, we can use the formula of Blasius

$$\lambda = \frac{0.3164}{\sqrt[4]{Re}}$$

② When $10^5 < Re < 3 \times 10^6$, we can use the formula of Nikuradse smooth pipe:

$$\lambda = 0.0032 + 0.221\, Re^{-0.237}$$

③ The more general formula is

$$\frac{1}{\sqrt{\lambda}} = 2\lg(Re\sqrt{\lambda}) - 0.8$$

Region IV: Turbulent transient region (the area between line cd and ef). In this case, λ is related with both Re and Δ/d, namely, $\lambda = f\left(Re, \frac{\Delta}{d}\right)$.

The formula for calculating λ in this region is as follows:

The Colebrook semi-empirical formula is commonly used to calculate λ in this region

$$\frac{1}{\sqrt{\lambda}} = 1.14 - 2\lg\left(\frac{\Delta}{d} + \frac{9.35}{Re\sqrt{\lambda}}\right)$$

The Colebrook formula is more complex, and it has a simplified form as follows:

$$\lambda = 0.11\left(\frac{\Delta}{d} + \frac{68}{Re}\right)^{0.25}$$

Region V: Turbulent flow in rough pipes (the area on the right of line ef). Experimental curve becomes a straight line parallel to horizontal axis. This means λ has nothing to do with Re, and only related to Δ/d, that is $\lambda = f\left(\frac{\Delta}{d}\right)$. It indicates that flow is in fully developed turbulent regime. The formula of Nikuradse rough-pipe region is commonly used for calculating λ as

$$\lambda = \frac{1}{\left[2\lg\left(3.7\frac{d}{\Delta}\right)\right]^2}$$

The significance of Nikuradse experiment lies in that it has entirely revealed the relationship among λ, Re and Δ/d in different regimes, and listed most empirical and semi-empirical formulas to determine λ in corresponding regions.

4.5.3 Moody Chart

Based on couples of experiment data, the relationship among λ, Re and Δ/d was plotted in 1940 by Moody into what was called the Moody chart for pipe friction shown in Fig. 4.11. The Moody chart is probably the most famous and useful figure in fluid mechanics. It can be used for circular and noncircular pipe flows and for open channel flow. The data can even be adapted as an approximation to boundary layer flows.

If Reynolds number Re and relative roughness Δ/d are known, then it is easy to find out the value of λ from the Moody chart. Table 4.1 gives the reference value of the surface roughness of the commonly used pipes.

Example 4.4 There is water flow with temperature $t = 6\ °C$ in a pipeline, and the dimensions of the pipeline are: diameter $d = 20$ cm, pipe length $l = 20$ m, and surface roughness $\Delta = 0.2$ mm. Try to determine the friction loss h_f if the flow rate is $Q = 24$ L/s.

Solution

When $t = 6\ °C$, the kinematic viscosity of water is $v = 0.0147\ \mathrm{cm^2/s}$

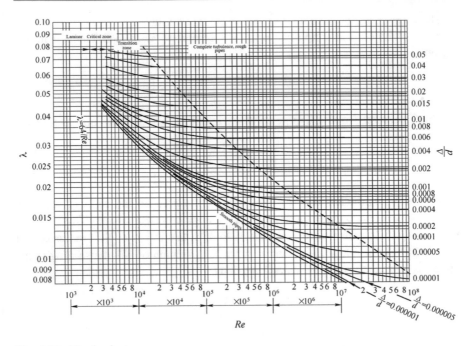

Fig. 4.11 Moody chart

Table 4.1 Surface roughness of common pipes

Pipe	Δ (mm)	Pipe	Δ (mm)
Clean brass or copper pipe	0.0015–0.002	Bituminous iron pipe	0.12
New seamless steel pipe	0.04–0.17	Galvanized iron pipe	0.15
New steel pipe	0.12	Glass, plastic pipe	0.001
Fine galvanized steel pipe	0.25	Rubber hose	0.01–0.03
Galvanized steel pipe	0.40	Wood pipe	0.25–1.25
Old rusty steel pipe	0.60	Concrete pipe	0.33
Common new cast iron pipe	0.25	Earthenware pipe	0.45–6.0
Old cast iron pipe	0.50–1.60		

The average velocity on cross section is

$$v = \frac{Q}{A} = \frac{24 \times 1000}{\frac{\pi}{4} \times 20^2} = 76.4 \text{ cm/s}$$

The Reynolds number is

$$\text{Re} = \frac{vd}{v} = \frac{76.4 \times 20}{0.0147} = 1.04 \times 10^5 > 2320 \text{ (in turbulent flow)}$$

The relative roughness is

$$\frac{\Delta}{d} = \frac{0.2}{20 \times 10} = 0.001$$

According to Re and $\frac{\Delta}{d}$, the friction factor can be obtained on Moody chart

$$\lambda = 0.027$$

The friction loss is

$$h_f = \frac{\lambda l}{d}\frac{v^2}{2g} = 0.027 \times \frac{20 \times 100}{20} \times \frac{76.4^2}{2 \times 980} = 8.04 \text{ cm H}_2\text{O}$$

Example 4.5 For water supply, petroleum supply and ventilation to a large equipment, the ambient temperature is 20 °C. The known data are listed in Table 4.2. Please calculate the friction loss h_f in water pipe, petroleum pipe and wind pipe respectively.

Solution

Showing the process of solving the problem in Table 4.3. First, the kinematic viscosity of water, petroleum and air at 20 °C can be obtained in Chap. 1 and listed in Table 4.3. Second, the surface roughness Δ of the pipe can be obtained from Table 4.1, and corresponding d/Δ can be calculated. Third, we can calculate the Reynolds number $Re = \frac{vd}{v} = \frac{4Q}{\pi d v}$ respectively.

In order to judge which region fluid belongs to, we need calculate the values of $22.2\left(\frac{d}{\Delta}\right)^{\frac{8}{7}}$ and $597\left(\frac{d}{\Delta}\right)^{\frac{9}{8}}$. According to the formulas in corresponding regions, we can calculate λ.

The friction loss h_f in different pipes can be calculated by the following formula.

$$h_f = \frac{\lambda l}{d}\frac{v^2}{2g} = \frac{8\lambda l Q^2}{\pi^2 d^5 g}$$

The results are listed in Table 4.3.

Table 4.2 Known data

	Water supply	Petroleum supply	Ventilation
Pipe	New cast iron pipe	Brass pipe	Seamless steel pipe
Pipe diameter d/cm	20	2	50
Pipe length l/m	20	10	10
Flow rate $Q/(\text{m}^3/\text{s})$	0.3	0.01	10

Table 4.3 Solution

	Water supply	Petroleum supply	Ventilation
$v/(\text{m}^2/\text{s})$	1.007×10^{-6}	8.4×10^{-6}	15.7×10^{-6}
Δ/mm	0.25	0.0018	0.10
d/Δ	800	11,111	5000
Re	1.90×10^{6}	7.58×10^{4}	1.62×10^{6}
$22.2\left(\frac{d}{\Delta}\right)^{\frac{8}{7}}$	46,150	9.33×10^{5}	3.75×10^{5}
$597\left(\frac{d}{\Delta}\right)^{\frac{9}{8}}$	1.10×10^{6}	–	8.66×10^{6}
Resistance region	V	III	IV
λ	0.0207	0.0104	0.0137
h_f	9.64 m H_2O	269 m (height of petroleum)	36.3 m (height of gas)

4.6   Calculation of Frictional Loss in Noncircular Duct

4.6.1   Using Darcy–Weisbach Equation

The characteristic length of circular section is its diameter d, and the characteristic length of the noncircular section is the hydraulic radius R. It is known that the relationship between them is $d = 4R$. Therefore, replacing d with $4R$, then for noncircular uniform flow, we have

$$h_f = \lambda \frac{l}{4R} \frac{v^2}{2g} \tag{4.34}$$

Calculation formulas can be dealt with as follows: replace d with $4R$, and replace Reynolds number of circular pipe $Re_{(d)} = \frac{vd}{v}$ with fourfold $Re_{(R)} = \frac{vR}{v}$ of noncircular pipe, then the formulas are suitable for noncircular pipe flow.

4.6.2   Using Chezy Equation

In order to apply Darcy–Weisbach equation to uniform flow in noncircular pipe, we transform the equation as

$$h_f = \frac{\lambda l}{d} \frac{v^2}{2g} = \frac{\lambda l}{4R} \frac{v^2}{2g} = \frac{l}{\frac{8g}{\lambda} R} \frac{1}{A^2} \frac{Q^2}{A^2} = \frac{Q^2 l}{c^2 R A^2}$$

Assuming $c^2 R A^2 = K^2$, then

$$h_f = \frac{Q^2 l}{K^2} \tag{4.35}$$

Thus, the formula of flow rate Q and velocity v are

$$Q = K\sqrt{\frac{h_f}{l}} = K\sqrt{i} = \sqrt{c^2 R A^2}\sqrt{i} \qquad (4.36)$$

$$v = c\sqrt{Ri} \qquad (4.37)$$

where i is friction loss per unit pipe length, $c = \sqrt{\frac{8g}{\lambda}}$ is Chezy coefficient, $K = cA\sqrt{R}$ is volumetric flow rate.

The above three equations were first proposed by Chezy, so it is called Chezy equation or Chezy formula, which has been widely used in pipeline and channel engineering calculation.

4.7 Theoretical Foundation of Boundary Layer

4.7.1 Basic Concept of Boundary Layer

In physics and fluid mechanics, a boundary layer is an important concept and refers to the layer of fluid in the immediate vicinity of a bounding surface where the effects of viscosity are significant [6–8].

The technique of boundary layer (BL) analysis can be used to compute viscous effects near solid walls and to "patch" these onto the outer inviscid motion. This patching is more successful as the body Reynolds number becomes larger, as shown in Fig. 4.12. In Fig. 4.12, a uniform stream u_0 moves parallel to a flat plate of length l. If the Reynolds number is low, the viscous region is very broad and extends far ahead and to the sides of the plate. The plate retards the oncoming stream greatly, and small changes in flow parameters cause large changes in the pressure distribution along the plate. Thus, although in principle it should be possible to patch the viscous and inviscid layers in a mathematical analysis, their interaction is strong and nonlinear. There is no existing simple theory for external

Fig. 4.12 Boundary layer structure

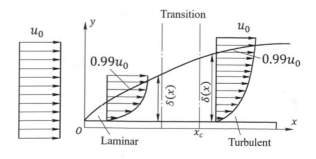

flow analysis. Such thick-shear-layer flows are typically studied by experiment or by numerical modeling of the flow field on a computer.

The flow layer in which the velocity increases from 0 to 0.99 u_0 is defined as boundary layer with the thickness δ.

There are three characteristics of boundary layer.

1. The thickness δ of boundary layer is much smaller than characteristic length l of object, $\delta \ll l, \frac{\delta}{l} \to 0$, namely, the thickness of boundary layer is extremely small.
2. The thickness of boundary layer increases along the flow direction of the flat plate.
   Because with increase of the flat plate length, friction loss also increases and fluid internal energy and flow velocity decrease. In order to meet the continuity requirement, the thickness of boundary layer must increase.
3. There exists laminar flow section, transition section and turbulent flow section in boundary layer. Under the transition section and turbulent flow section, there also exists a bottom layer of laminar flow.

4.7.2 Thickness of Flat Plate Boundary Layer

In the boundary layer, the position of the change from laminar flow to turbulent flow is called transitional point x_c, and the corresponding Reynolds number is called critical Reynolds number Re_c which is related to turbulent intensity and wall roughness. The value of critical Reynolds number can be obtained by experiment shown as follows.

$$Re_c = \frac{u_0 x_c}{v} = 3.0 \times 10^5 - 3.0 \times 10^6 \tag{4.38}$$

When the flat plate is long, the length of laminar boundary layer and the transition section are small by comparison with that of turbulent boundary layer. The thickness of laminar boundary layer confirmed by theoretical analysis and experiments is

$$\frac{\delta}{x} = \frac{5}{\sqrt{Re_x}} \tag{4.39}$$

The thickness of turbulent boundary layer is

$$\frac{\delta}{x} = \frac{0.37}{(Re_x)^{0.2}} \tag{4.40}$$

4.7.3 Boundary Layer Separation

When viscous fluid flows over curved bodies, velocity along curved boundary layer will change, and the flow resistance has to vary a lot at the separated point. Calculation for curved boundary layer is very complicated and we will not discuss it here. In this section, we will mainly focus on the illustration of curved boundary layer separation.

In actual engineering, the objects usually have shapes of curved surface (streamlined or non-streamlined object). When a fluid flows over non-streamlined object in many cases, the boundary layer on the surface will separate from the surface at a certain position, and reflux in the reverse direction of the mainstream occurs in vicinity of surface. This kind of phenomenon is called boundary layer separation in fluid mechanics, as shown in Fig. 4.13a. For streamlined object, boundary layer separation may also happen in abnormal condition, as shown in Fig. 4.13b.

In addition, if the streamline of fluid flow change suddenly because of sudden expansion, sudden contraction, bending pipe and so on in Fig. 4.14, the separation of the boundary layer will appear due to inertia and generate vortex region.

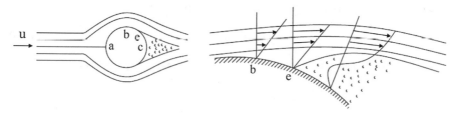

Fig. 4.13 Boundary layer separation

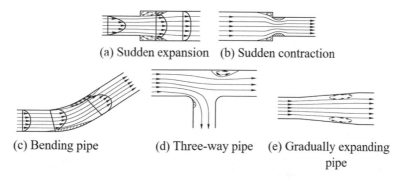

(a) Sudden expansion (b) Sudden contraction

(c) Bending pipe (d) Three-way pipe (e) Gradually expanding pipe

Fig. 4.14 Vortex region generated by sudden change of streamline

4.8 Minor Head Loss in Pipeline

4.8.1 Minor Head Loss of Sudden Expansion

Figure 4.15 shows a sudden flow expansion in a pipe. Two diameters of the pipe
are d_1 and d_2, respectively. Water flows from the smaller diameter pipe into the
larger diameter pipe leading to a recirculation zone generation, and the length of the
recirculation region is (5–8) d_2. Since l is short enough, the frictional loss h_f can be
ignored comparing with minor loss.

Write Bernoulli's equation of total flow between cross section 1-1 and 2-2

$$z_1 + \frac{p_1}{\gamma} + \frac{\alpha_1 v_1^2}{2g} = z_2 + \frac{p_2}{\gamma} + \frac{\alpha_2 v_2^2}{2g} + h_r \tag{4.41}$$

The momentum equation of control volume between cross section A-A and 2-2 is

$$\sum F = p_1 A_1 + P + G \sin\theta - p_2 A_2 = \rho Q(\alpha_{02} v_2 - \alpha_{01} v_1) \tag{4.42}$$

where P is the reaction force exerting on control volume by the pipe wall of annular
area $A_2 - A_1$ at section $A - A$.

According to the relationship between pressure and force, we have

$$P = p_1(A_2 - A_1) \tag{4.43}$$

Showing in Fig. 4.15, the projection of gravity G along the pipe center line is

$$G \sin\theta = \gamma A_2 l \frac{z_1 - z_2}{l} = \gamma A_2(z_1 - z_2) \tag{4.44}$$

Substituting Eqs. (4.43) and (4.44) and continuity equation $Q = A_1 v_1 = A_2 v_2$
into Eq. (4.42), we have

Fig. 4.15 Sudden flow
expansion

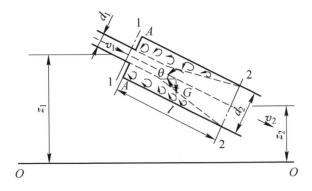

$$(z_1 - z_2) + (\frac{p_1}{\gamma} - \frac{p_2}{\gamma}) = \frac{(\alpha_{02}v_2 - \alpha_{01}v_1)v_2}{g}$$

Substituting the above equation into Eq. (4.41), we have

$$h_r = \frac{(\alpha_{02}v_2 - \alpha_{01}v_1)v_2}{g} + \frac{\alpha_1 v_1^2 - \alpha_2 v_2^2}{2g}$$

When Reynolds number is large enough, $\alpha_1, \alpha_2, \alpha_{01}$ and α_{02} are all close to 1, so the above equation can be rewritten as

$$h_r = \frac{(v_1 - v_2)^2}{2g} \tag{4.45}$$

Substituting $v_2 = A_1 v_1/A_2$ and $v_1 = A_2 v_2/A_1$ into the above equation respectively, we have

$$h_r = \left(1 - \frac{A_1}{A_2}\right)^2 \frac{v_1^2}{2g} = \zeta_1 \frac{v_1^2}{2g} \tag{4.46}$$

$$h_r = \left(\frac{A_2}{A_1} - 1\right)^2 \frac{v_2^2}{2g} = \zeta_2 \frac{v_2^2}{2g} \tag{4.47}$$

The above equation is named Borda equation, where ζ_1 and ζ_2 are the minor (local) loss coefficient or factor with velocity v_1 and v_2 respectively.

4.8.2   Other Types of Minor Loss

In terms of above analysis, minor loss can be expressed as the product of flow velocity head and a coefficient, that is

$$h_r = \zeta \frac{v^2}{2g} \tag{4.48}$$

Minor loss coefficient ζ varies with different local devices. If the local device is fixed in an equal-radius pipe, there exists only one value of minor loss coefficient for head loss calculation with average velocity. If the local device is fixed just at the changing position of a pipe, it would have two values of minor loss factor. Usually we select the parameters of latter part to calculate the minor loss such as latter velocity head and minor loss coefficient.

Minor loss coefficients of several common devices can be determined as follows:

(1) A sudden contraction of the pipe diameter (Fig. 4.16). The value of ζ varies with the area ratio of A_2/A_1, as shown in Table 4.4.

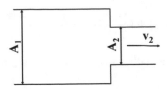

Fig. 4.16 Sudden contraction of the pipe diameter

Table 4.4 The minor loss coefficients for sudden contraction of the pipe diameter

A_2/A_1	0.01	0.1	0.2	0.3	0.4	0.5	0.6	0.7	0.8	0.9	1
ζ		0.490	0.469	0.431	0.387	0.343	0.298	0.257	0.212	0.161	0.070	0

(2) Gradually expanding pipe (Fig. 4.17).
The value of ζ can be determined by the following formula:

$$\zeta = \frac{\lambda}{8\sin\frac{\alpha}{2}}\left[1 - \left(\frac{A_1}{A_2}\right)^2\right] + K\left(1 - \frac{A_1}{A_2}\right), \tag{4.49}$$

where K is related to expansion angle α, and its value is listed in Table 4.5 when $\frac{A_1}{A_2} = \frac{1}{4}$.

(3) Gradual contraction pipe (Fig. 4.18). The value of ζ can be calculated by the following formula:

$$\zeta = \frac{\lambda}{8\sin\frac{\alpha}{2}}\left[1 - \left(\frac{A_2}{A_1}\right)^2\right] \tag{4.50}$$

(4) Bending pipe (elbow) (Fig. 4.19) and Folding pipe (Fig. 4.20).
The value of ζ for elbow can be calculated by

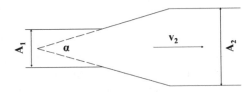

Fig. 4.17 Gradually expanding pipe

Table 4.5 The values of K for calculating ζ of gradually expanding pipe

α	2°	4°	6°	8°	10°	12°	14°	16°	20°	25°
K	0.022	0.048	0.072	0.103	0.138	0.177	0.221	0.270	0.386	0.645

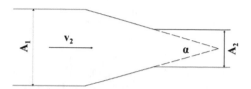

Fig. 4.18 Gradual contraction pipe

Fig. 4.19 Bending pipe

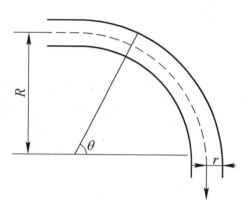

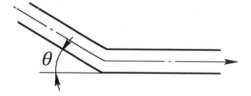

Fig. 4.20 Folding pipe

$$\zeta = \left[0.131 + 1.847\left(\frac{r}{R}\right)^{3.5}\right]\frac{\theta}{90°} \qquad (4.51)$$

When $\theta = 90°$, the minor loss coefficients of common elbows are listed in Table 4.6.

For cast iron bending pipe $\frac{r}{R} = 0.75$, its minor loss coefficient is $\zeta = 0.9$.

The value of ζ for folding pipe can be calculated by

Table 4.6 The minor loss coefficients of $\theta = 90°$ elbow

$\frac{r}{R}$	0.1	0.2	0.3	0.4	0.5	0.6	0.7	0.8	0.9	1
ζ	0.132	0.138	0.158	0.206	0.294	0.440	0.661	0.977	1.408	1.978

$$\zeta = 0.946 \sin^2 \left(\frac{\theta}{2}\right) + 2.407 \sin^4 \left(\frac{\theta}{2}\right) \tag{4.52}$$

The minor loss coefficients of folding pipes are listed in Table 4.7.

(5) Three-way pipe The minor loss coefficients are listed in Table 4.8.

(6) Gate valve (Fig. 4.21) and Cut-off valve (Fig. 4.22). The minor loss coefficient varies according to the opening degree, and the values of ζ are listed in Table 4.9.

(7) Inlet, outlet, and other common fittings for pipe. The values of ζ are listed in Table 4.10.

Table 4.7 The minor loss coefficients of folding pipe

θ	20°	40°	60°	80°	90°	100°	110°	120°	130°	160°
ζ	0.046	0.139	0.364	0.741	0.985	1.260	1.560	1.861	2.150	2.431

Table 4.8 The minor loss coefficients of three-way pipes

90° three-way pipe				
ζ	0.1	1.3	1.3	3
45° three-way pipe				
ζ	0.15	0.05	0.5	3

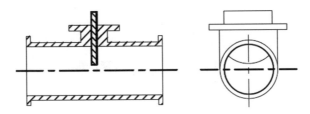

Fig. 4.21 Gate valve

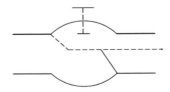

Fig. 4.22 Cut-off valve

Table 4.9 The minor loss coefficients of gate valve and cut-off valve

Opening degree/%	10	20	30	40	50	60	70	80	90	100
Gate valve ζ	60	15	6.5	3.2	1.8	1.1	0.60	0.30	0.18	0.1
Cut-off valve ζ	85	24	12	7.5	5.7	4.8	4.4	4.1	4.0	3.9

4.8.3 Aggregation of Head Loss

The above minor loss coefficients are measured without the interference of other resistances, otherwise, their values will change.

The complex effect of actual installation situation is unknown, therefore, to calculate the total head (pressure, energy) loss is just to add all frictional losses and minor losses together, namely aggregation of head loss.

According to aggregation of head loss, the total head loss of a pipe can be expressed as

$$h_l = h_f + \sum h_r = \left(\lambda \frac{l}{d} + \sum \zeta \right) \frac{v^2}{2g} \qquad (4.53)$$

To facilitate calculation, we can transfer the minor loss factor into equivalent length as follows:

$$\zeta = \lambda \frac{l_e}{d} \quad \text{or} \quad l_e = \frac{\zeta}{\lambda} d, \qquad (4.54)$$

where l_e is called the equivalent length of minor loss. So the total head loss can be simplified as

$$h_l = \lambda \frac{l + \sum l_e}{d} \frac{v^2}{2g} = \lambda \frac{L}{d} \frac{v^2}{2g}, \qquad (4.55)$$

where $L = l + \sum l_e$ is called the total drag length of the pipe. The equivalent lengths of various common local devices can be referred in professional manuals.

Actual pipelines are mostly composed of several equal-radius pipes and local devices, so the head loss can be calculated by the following formula:

$$h_l = \sum h_f + \sum h_r$$
$$h_l = \sum_{i=1}^{n} \frac{\lambda_i l_i}{d_i} \frac{v_i^2}{2g} + \sum_{j=1}^{m} \zeta_j \frac{v_j^2}{2g} = \sum_{i=1}^{n} \frac{\lambda_i l_i}{d_i} \frac{v_i^2}{2g} + \sum_{j=1}^{m} \frac{\lambda_j l_{ej}}{d_j} \frac{v_j^2}{2g} \qquad (4.56)$$

Example 4.6 There is a pipeline from higher reservoir to lower reservoir, as shown in Fig. 4.23. It is known that height difference of two reservoirs is $H = 40$ m, pipe length $L = 200$ m, pipe diameter $d = 50$ mm, and $\theta = 90°$ elbow $r/R = 0.5$. The pipe is galvanized steel pipe ($\Delta = 0.4$ mm). How much water this pipeline can supply for day and night when the water temperature is 20 °C?

Table 4.10 The minor loss coefficients of inlet, outlet and other common fittings for pipe

Fitting	Diagram	Coefficient	Fitting	Diagram	Coefficient
Sharp inlet		$\zeta = 0.5$	Curved inlet		$\zeta = 0.2$
Sharp inclined inlet		$\zeta = 0.505 + 0.303 \sin\theta + 0.226 \sin^2\theta$	Pipe outlet		$\zeta = 1$
Gate		$\zeta = 0.12$ (fully opened)	Butterfly valve		When $\alpha = 20°$, $\zeta = 1.54$; when $\alpha = 45°$, $\zeta = 18.7$
Cyclone separator		$\zeta = 2.5$–3.0	Screen filter (with bottom valve)		$\zeta = 10$; without bottom valve, $\zeta = 5$–6
Check valve (one-way valve)		$\zeta = 1.7$–14 varying with opening degree	Gradual contraction short pipe ($\alpha = 5°$)		$\zeta = 0.06$; for nozzle, $\zeta = 0.06$

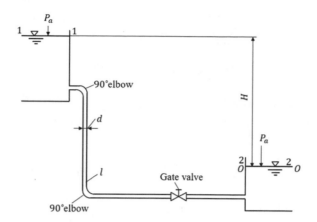

Fig. 4.23 Water supply pipeline

Solution

According to Table 1.2, when $t = 20\ °C$, the kinematic viscosity of water is $v = 1.007 \times 10^{-6}\ \mathrm{m^2/s}$.

Take water surface O-O of lower reservoir as datum reference, and the Bernoulli equation between cross section 1-1 and 2-2 is

$$H + \frac{p_a}{\gamma} + \frac{\alpha_1 v_1^2}{2g} = 0 + \frac{p_a}{\gamma} + \frac{\alpha_2 v_2^2}{2g} + \frac{\lambda l}{d}\frac{v^2}{2g} + \sum \zeta \frac{v^2}{2g} \qquad (4.57)$$

According to given conditions, $v_1 = v_2 \approx 0$
Minor loss coefficient of pipe inlet is $\zeta_1 = 0.5$
90° elbows: $\zeta_2 = 0.294 \times 2 = 0.588$
Gate valve (fully opened): $\zeta_3 = 0.1$
Pipe outlet: $\zeta_4 = 1.0$
Therefore, $\sum \zeta = \zeta_1 + \zeta_2 + \zeta_3 + \zeta_4 = 2.188$
Substituting above equation into Eq. (4.57), we have

$$H = \left(\frac{\lambda l}{d} + 2.188\right)\frac{v^2}{2g}$$
$$= (4000\lambda + 2.188)\frac{v^2}{2g} \qquad (4.58)$$

The relative roughness of the pipe is $\frac{\Delta}{d} = \frac{0.4}{50} = 0.008$. Assuming that the water flows in region IV, we can take $\lambda = 0.036$ temporarily on Moody chart, and substitute it into Eq. (4.58), then

$$v = \sqrt{\frac{2 \times 9.8 \times 40}{4000 \times 0.036 + 2.188}} = 2.316 \text{ m/s}$$

$$Re = \frac{vd}{v} = \frac{2.316 \times 0.05}{0.01007 \times 10^{-4}} = 1.15 \times 10^5$$

It can be seen from Moody chart according to $\frac{\Delta}{d}$ and Re that the water flow is indeed in transient region, and it is also reasonable to take $\lambda = 0.036$.

The flow rate in the pipe is

$$Q = Av = \frac{\pi}{4} \times 0.05^2 \times 2.316 = 0.00455 \text{ m}^3/\text{s}$$

The water supply for day and night is

$$V = 24 \times 3600Q = 24 \times 3600 \times 0.00455 = 393.12 \text{ m}^3.$$

4.9 Problems

4.1. There is water flow in a pipe with diameter 10 mm. The average velocity on cross section is 0.25 m/s, and the water temperature is 10 °C. Try to determine the flow pattern of water. If the diameter is changed to 25 mm. The average velocity and water temperature do not change. Try to determine the corresponding flow pattern. When the diameter is still 25 mm with same water temperature, determining the flow rate in order to make flow regime change from turbulent to laminar flow.

4.2. As shown in Fig. 4.24, there is a trapezoid cross-section drainage ditch with width $b = 70$ cm and slope 1 : 1.5. When the water depth $h = 40$ cm, average velocity on cross section $v = 5.0$ cm/s, and water temperature 10 °C, determining the flow regime of water at this time. If the water depth and temperature remain the same, determining the average velocity on cross section to change the flow regime.

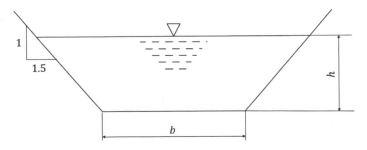

Fig. 4.24 Problem 4.2

4.3. As shown in Fig. 4.25, there is oil flow with relative density 0.9 in a horizontal pipe with diameter $d = 5$ cm and length $L = 6$ m. The reading of mercury manometer is $h = 13.5$ cm. The outflow of oil in three minutes is 5000 N. Try to determine the dynamic viscosity μ of oil.

4.4. There is oil flow with kinematic viscosity $v = 0.2\,\text{cm}^2/\text{s}$ in a pipe, and the average velocity of oil is $v = 1.5$ m/s. The friction loss every 100 m length is 40 cm. Try to determine the relationship between friction factor and Reynolds number.

4.5. There is oil flow with relative density 0.85 and kinematic viscosity $v = 0.125\,\text{cm}^2/\text{s}$ in a seamless steel pipe $(\Delta = 0.04\,\text{mm})$. The pipe diameter is $d = 30$ cm, and the flow rate is $Q = 0.1\,\text{m}^3/\text{s}$. Try to judge flow pattern and determine:

(1) friction factor λ
(2) thickness δ of laminar sublayer
(3) shear stress τ_0 on pipe wall.

4.6. As shown in Fig. 4.26, water flows in a vertical pipe with diameter d and length l into the atmosphere. The water height in tank is h. The minor head loss of the piping system can be ignored, and the friction factor is λ.

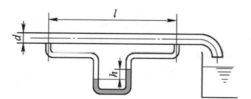

Fig. 4.25 Problem 4.3

Fig. 4.26 Problem 4.6

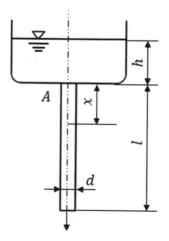

(1) try to determine pressure on cross section A at inlet of pipe.
(2) what is the value of h to make the pressure at point A equal to atmospheric pressure?
(3) determine the average velocity in the pipe.
(4) what is the value of h to make the flow rate irrelevant to the length l of pipe?
(5) if $d = 4$ cm, $l = 5$ m, $h = 1$ m, $\lambda = 0.04$, determining the pressure at point A (that is, $x = 0$) and $x = 1, 2, 3, 4$ m respectively.

4.7 There is water flow in a pipe with diameter $d = 10$ cm and surface roughness $\Delta = 0.3$ mm. The water temperature is $5\,^\circ$C and the average velocity is $v = 1.5$ m/s.
(1) which region does the water flow belong to?
(2) what is the value of λ?

4.8. The crude oil with temperature $20\,^\circ$C (kinematic viscosity $v = 7.2$ mm^2/s) passes through a new cast iron pipe ($\Delta = 0.24$ mm) with length 800 m and diameter 300 mm. Only friction loss is taken into consideration. How much pressure is required when the flow rate is 0.25 m^3/s?

4.9. There is petroleum flow in a pipe with diameter $d = 20$ cm and length $l = 1000$ m. The flow rate is $Q = 40$ L/s and the kinematic viscosity of petroleum is $v = 1.6$ cm^2/s. Try to determine friction loss h_f.

4.10. A pipe with length $l = 1000$ m and diameter $d = 150$ mm is used to transport crude oil. When the flow rate is $Q = 40$ L/s, oil temperature is $t = 38\,^\circ$C and corresponding kinematic viscosity is $v = 0.3$ cm^2/s, the output power of oil pump is $N = 7.35$ kW; If the temperature decreases to $t = -1\,^\circ$C and $v = 3$ cm^2/s, how much power is required in order to maintain same flow rate? (assuming that the specific weight of crude oil is $\gamma = 8.829$ kN/m^3 and does not change with temperature).

4.11. There is a rectangular air duct with cross-section area 1200 mm × 600 mm, surface roughness $\Delta = 0.1$ mm and length $l = 12$ m. The air temperature is $45\,^\circ$C and the flow rate is 42,000 m^3/h. The reading in an alcohol manometer with inclined angle 30° is $a = 7.5$ mm. The alcohol density is $\rho = 860$ kg/m^3. Determine friction factor λ of the air duct and compare it with the value on Moody chart.

4.12. As shown in Fig. 4.27, there is a sudden expansion pipe. ΔH is the difference of pressure head between two piezometers. If the flow velocity of the large and small pipe remain unchanged, determining the ratio of large pipe diameter D to small pipe diameter d in order to obtain maximum value of ΔH and expressing the maximum value ΔH_{max} by velocity of small pipe.

4.13. The horizontal pipe diameter is suddenly enlarged from $d = 100$ mm to $D = 150$ mm, and water flow rate is $Q = 2$ m^3/min, as shown in Fig. 4.28.

Fig. 4.27 Problem 4.12

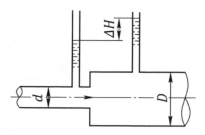

Fig. 4.28 Problem 4.13

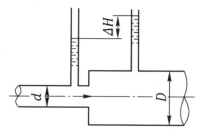

(1) try to find the minor head loss of sudden expansion pipe.
(2) try to find the difference ΔH of pressure head between two piezometers.
(3) if the pipe is gradually expanding pipe and minor head loss can be ignored, determining the corresponding ΔH.

4.14. As shown in Fig. 4.29, there is a horizontal sudden contraction pipe with $d_1 = 15$ cm and $d_2 = 10$ cm. The flow rate of water is $Q = 2 \, \text{m}^3/\text{min}$, and the reading in a mercury manometer is $h = 8$ cm. Find the minor head loss of sudden contraction pipe.

4.15. As shown in Fig. 4.30, there is water flow in a pipe with diameter $d = 50$ mm and friction factor $\lambda = 0.0285$, and the flow rate is $15 \, \text{m}^3/\text{h}$. The pipe length between point A and B is 0.8 m. The reading in a mercury manometer is $\Delta h = 20$ mm. Determine the minor loss coefficient of bending pipe.

4.16. As shown in Fig. 4.31, there is a fire hose with diameter $d_1 = 20$ mm and length $l = 18$ m.

Fig. 4.29 Problem 4.14

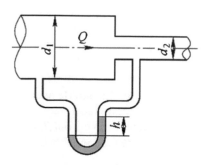

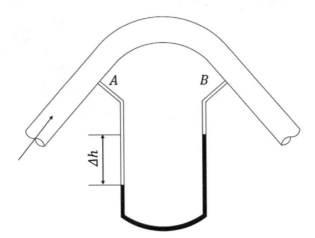

Fig. 4.30 Problem 4.15

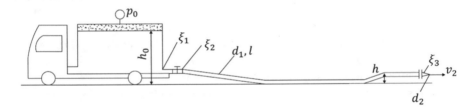

Fig. 4.31 Problem 4.16

The diameter of nozzle at outlet is $d_2 = 3$ mm. The minor loss coefficients of the fittings in the pipe are: inlet $\zeta_1 = 0.5$, valve $\zeta_2 = 3.5$ and nozzle $\zeta_3 = 0.1$ (relative to the outflow velocity of nozzle). The friction factor is $\lambda = 0.03$. The relative pressure on water surface in the tank is $p_0 = 4 \times 10^5$ Pa, and $h_0 = 3$ m, $h = 1$ m. Determine the outflow velocity of nozzle.

4.17. In order to determine the minor loss coefficient ζ of 90° elbow, we can use the device as shown in Fig. 4.32. It is known that AB pipe length is 10 cm and the diameter is 50 mm. For turbulent flow in rough pipes, the friction factor λ is 0.03. The flow rate is 2.74 L/s currently, and the height difference between piezometer 1 and 2 is 62.9 cm. Try to determine the minor loss coefficient ζ of elbow.

4.18. As shown in Fig. 4.33, the pipe diameter is $d = 25$ mm. $l_1 = 8$ m, $l_2 = 1$ m, $H = 5$ m. The nozzle diameter is $d_0 = 10$ mm, and the minor loss coefficients of inlet and elbow are $\zeta_1 = 0.5$ and $\zeta_2 = 0.1$ respectively. For nozzle, $\zeta_3 = 0.1$ (relative to the outflow velocity of nozzle). The friction factor is $\lambda = 0.03$. Try to determine jet height h.

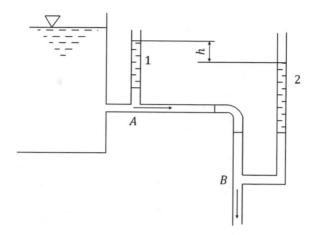

Fig. 4.32 Problem 4.17

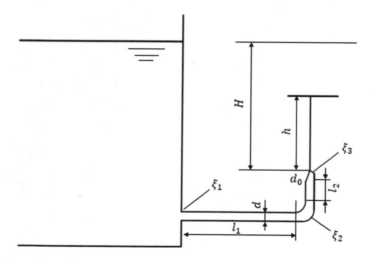

Fig. 4.33 Problem 4.18

4.19. Reservoir A and B are connected by two pipes with different diameter, as shown in Fig. 4.34. It is known that $d_1 = 200\,\text{mm}, l_1 = 15\,\text{m}, l_2 = 20\,\text{m}$, surface roughness $\Delta = 0.8\,\text{mm}$ and $H = 20\,$m. There are two 90° elbows with $\frac{d}{r} = 0.5$ and fully opened gate valve in the pipeline. The water temperature is $t = 20\,$°C. Find the flow rate in the pipeline.

4.20. As shown in Fig. 4.35, passing through a funnel with diameter $d_2 = 50$ mm, height $h = 40$ cm and minor loss coefficient $\zeta = 0.25$, gasoline goes into a tank. The gasoline flows from a higher oil storage tank to the funnel through a short pipe, and the pipe diameter is $d_1 = 30$ mm. The minor loss coefficients of

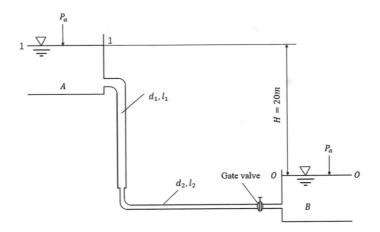

Fig. 4.34 Problem 4.19

Fig. 4.35 Problem 4.20

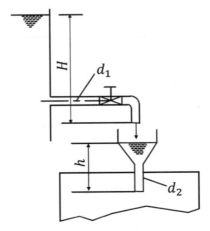

valve and elbow are $\zeta = 0.85$ and $\zeta = 0.8$ respectively. For inlet of the pipe, $\zeta = 0.5$. The friction loss can be ignored. Try to determine the height H in the higher tank in order to make sure the gasoline doesn't flow over the edge of the funnel, and find corresponding flow rate that flows into the tank.

4.21. As shown in Fig. 4.36, there are two reservoirs connected by a pipe. The pipe diameter is $d = 500$ mm, and the length is 100 m. The surface roughness of pipe wall is 0.6 mm. The opening degree of gate valve is 60% and the dimension of 90° elbow is $R = 2d$. The water temperature is 20 °C and the flow rate is 0.5 m³/s. The water surface of two reservoirs remains unchanged. Find the height difference H.

4.22. As shown in Fig. 4.37, water flows out from a reservoir. The pipe diameter is $d = 15$ cm. $l_1 = 30$ m, $l_2 = 60$ m and $H_2 = 15$ m. It is known that friction

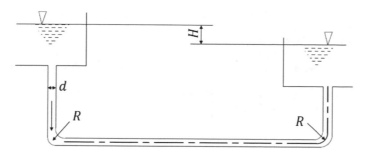

Fig. 4.36 Problem 4.21

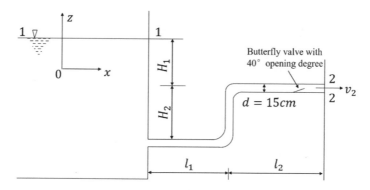

Fig. 4.37 Problem 4.22

factor of the pipe is $\lambda = 0.023$, and the minor loss coefficients of elbow and butterfly valve with 40° opening degree are $\zeta_1 = 0.9$ and $\zeta_2 = 10.8$ respectively. For inlet of the pipe, $\zeta = 0.5$. Try to determine:

(1) the flow rate Q of the pipe when $H_1 = 10$ m;
(2) the value of H_1 if the flow rate in the pipe is 0.06 m³/s.

References

1. White, F.M.: Fluid Mechanics, 7th edn. McGraw-Hill, New York (2011)
2. Fox, R.W., McDonald, A.T., Pritchard, P.J.: Introduction to Fluid Mechanics, vol. 5, 8th edn. Wiley, New York (2011)
3. Reynolds, O.: An experimental investigation of the circumstances which determine whether the motion of water shall be direct sinuous and of the law or resistance in parallel channels. Philos. Trans. R. Soc. **174**, 935–982 (1883)
4. Xie, Z.: Engineering Fluid Mechanics, 4th edn. Metallurgical Industry Press, Beijing (2014)

5. Song, H.: Engineering Fluid Mechanics and Environmental Application. Metallurgical Industry Press, Beijing (2016)
6. Schlichting, H.: Boundary Layer Theory, 8th edn. McGraw-Hill, New York (2000)
7. Prandtl, L.: Über Flüßigkeitsbewegung bei sehr kleiner Reibung. Verh. 3. Intern. Math. Kongr., Heidelberg (1904), pp. 484–491, Nachdruck: Ges. Abh., pp. 575–584. Springer, Heidelberg (1961)
8. von Kármán, T.: Über laminare und turbulente Reibung. Zeitschrift für angewandte Mathematik und Mechanik 1, 233–252 (1921)

Pipe Network and Orifice, Nozzle Flow

5

Abstract

For pipe flow calculation, the piping system can be divided into two parts, that is single pipe flow and multiple pipes flow. The combination of multiple pipes can make up the pipe network. The researches on orifice and nozzle flow are also meaningful in engineering application. In this chapter, we will first introduce single pipe flow and multiple pipes flow. For multiple pipes flow, we mainly discuss series flow, parallel flow and uniformly variable mass outflow. Then we will figure out characteristics of orifice flow and nozzle flow. For them, we will pay more attention to their discharge coefficients for velocity and flow rate.

Keywords

Series flow · Parallel flow · Orifice flow · Nozzle flow · Discharge coefficient

5.1 Flow in Single Pipe

Single pipe always involves one pipe with constant diameter. The hydraulic calculation in single pipe is the simplest among pipeline hydraulic calculation, which is a fundamental for the hydraulic calculation in multiple pipes. The flow in single pipe can be classified into short pipe and long pipe flow. For hydraulic calculation in a short pipe, we cannot ignore the minor head loss, otherwise it can be ignored in a long pipe.

5.1.1   Hydraulic Calculation in a Short Pipe

The suction pipe of a pump and siphon all belong to the short pipes, so their minor head loss cannot be neglected in the calculation. The following is an example for short pipe hydraulic calculation.

Example 5.1 As shown in Fig. 5.1, there is a pump installed in a piping system. The diameter of cast iron pipe is $d = 150$ mm, and the length is $l = 180$ m. The minor loss coefficient of screen filter without bottom valve is $\zeta_1 = 6$, for inlet and outlet of the pipe $\zeta_2 = 0.5$, $\zeta_5 = 1$, for three elbows $\zeta_3 = 0.294 \times 3$, for valve $\zeta_4 = 3.9$. Assuming that the height $h = 100$ m, the flow rate $Q = 225$ m^3/h and the temperature of water $t = 20$ °C, find output power of the pump.
Solution

When $t = 20$ °C, the kinematic viscosity of water $v = 1.007 \times 10^{-6}$ m^2/s (Table 1.2), then

$$\text{Re} = \frac{vd}{v} = \frac{4Q}{\pi d v} = \frac{4 \times 225}{3600\pi \times 0.15 \times 1.007 \times 10^{-6}} = 5.25 \times 10^5$$

Cast iron pipe $\Delta = 0.30$ mm, $\frac{\Delta}{d} = 0.002$

Check the Moody chart according to Re and Δ/d, $\lambda = 0.024$.

According to the given conditions, $\sum \zeta = \zeta_1 + \zeta_2 + \zeta_3 + \zeta_4 + \zeta_5 = 12.28$, then the equivalent pipe length of minor head loss is

$$\sum l_e = \frac{\sum \zeta}{\lambda} d = \frac{12.28}{0.024} \times 0.15 = 76.75 \text{ m}$$

Substituting the equation $v = \frac{4Q}{\pi d^2}$ into $h_l = \lambda \frac{l + \sum l_e}{d} \frac{v^2}{2g}$, we have

Fig. 5.1   A piping system [1]

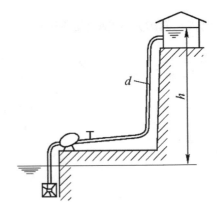

$$h_l = \frac{8\lambda(l + \sum l_e)Q^2}{g\pi d^5} = \frac{8 \times 0.024 \times (180 + 76.75) \times 225^2}{9.8 \times \pi^2 \times 0.15^5 \times 3600^2} = 26.22 \text{ m}$$

Thus, the pump's head rise is

$$H = h + h_l = 100 + 26.22 = 126.22 \text{ m}$$

The output power of the pump is

$$P = \gamma QH = 9800 \times \frac{225}{3600} \times 126.22 = 77310 \text{ W} = 77.3 \text{ kW}$$

5.1.2 Hydraulic Calculation in a Long Pipe

We take an example to analyze hydraulic characteristics and calculation method for long pipe flow. As shown in Fig. 5.2, water is drained away from a water tank by a pipe with diameter d, length l and height H from water surface to the outlet.

With 0–0 as the datum reference, then Bernoulli equation between section 1–1 and 2–2 is:

$$H + \frac{p_a}{\gamma} + \frac{\alpha_1 v_1^2}{2g} = 0 + \frac{p_a}{\gamma} + \frac{\alpha_2 v_2^2}{2g} + h_l$$

Because of long pipe, we neglect minor head loss h_r and velocity head $\frac{\alpha_2 v_2^2}{2g}$, then we have $h_l = h_f$. Assuming that $v_1 \approx 0$, we have

$$H = h_f \tag{5.1}$$

The above equation shows that the total head of long pipe equals frictional loss, and the frictional loss of long pipe flow can be calculated by Chezy formula, namely

$$h_f = \frac{Q^2 l}{K^2}, \tag{5.2}$$

Fig. 5.2 Long pipe flow

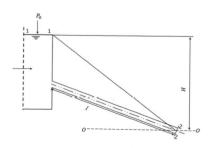

where Q is the flow rate, l is the length of pipe, and K is volumetric flow rate which can be referred from textbooks or professional manuals.

The Chezy formula is generally adopted in hydraulic calculation of long pipe flow. K can be expressed as follows:

$$K = cA\sqrt{R} = \sqrt{\frac{8g}{\lambda}} \times \frac{1}{4}\pi d^2 \sqrt{\frac{d}{4}} = 3.462\sqrt{\frac{d^5}{\lambda}}\ \mathrm{m^3/s}$$

The relationship between frictional factor λ and Chezy factor c can be expressed as follows.

$$\lambda = \frac{8g}{c^2}$$

The common empirical formula for Chezy factor c calculation is Manning formula, that is

$$c = \frac{1}{n}R^{\frac{1}{6}}, \tag{5.3}$$

where R is hydraulic radius.

n is the roughness coefficient which has different values for different types of boundary roughness. In engineering, the approximate values for different n can be referred from Table 5.1.

K is the function of d and n. Table 5.2 gives the values of K with different d and n.

Example 5.2 Assuming that the flow rate of a pipe $Q = 250$ L/s, the length $l = 2500$ m, the head $H = 30$ m. The material of the pipe is new cast iron. Find the diameter of the pipe.

Table 5.1 Values for roughness coefficient [1]

Number	Pipe types	n
1	New and clean cast iron or steel pipe	0.0111
2	Concrete or reinforced concrete pipe	0.0125
3	Welding metal pipe	0.012
4	Riveted metal pipe	0.013
5	Wood pipe with large diameter	0.013
6	Non-lined pressure pipe in rocks	0.025–0.04
7	Dirty water-supply or drainage pipe	0.014

Table 5.2 Values for K with d and n

Diameter d (mm)	$K/\text{L s}^{-1}$		
	$n = 0.0111, \frac{1}{n} = 90$ Clean cast iron pipe	$n = 0.0125, \frac{1}{n} = 80$ Normal cast iron pipe	$n = 0.0143, \frac{1}{n} = 70$ Dirty cast iron pipe
50	9.624	8.46	7.043
75	28.31	24.94	21.83
100	61.11	53.72	47.01
125	110.8	97.4	85.23
150	180.2	158.4	138.6
200	388.0	341.0	298.5
250	703.5	618.5	541.2
300	1144	1006	880
350	1727	1517	1327
400	2464	2166	1895
450	3373	2965	2594
500	4467	3927	3436
600	7264	6386	5587
700	10,960	9632	8428
800	15,640	13,750	12,030
900	21,420	18,830	16,470
1000	28,360	24,930	21,820

Solution

$$K = \frac{Q}{\sqrt{\frac{H}{l}}} = \frac{250}{\sqrt{\frac{30}{2500}}} = 2283 \text{ L/s}$$

According to Table 5.2, when $n = 0.0111$ and $K = 2283$ L/s, the diameter of pipe is between 350 and 400 mm, and can be calculated as follows:

$$d = 350 + \frac{400 - 350}{2464 - 1727} \times (2283 - 1727) = 378 \text{ mm}$$

5.2 Flow in Multiple Pipes

There are multiple pipes in actual pipe systems. The multiple pipe systems include series flow, parallel flow, variable mass outflow and so on, where only the friction losses are taken into consideration for hydraulic calculation.

5.2.1 Series Flow

As shown in Fig. 5.3, the flow in pipe system which consists of pipes with different diameters is series flow. The flow rate in each pipe may be constant, or variable due to the leakage in the end of each pipe.

We assume the length, the diameter, and flow rate of each pipe are l_i, d_i, and Q_i respectively. The flow rate leakage in each pipe is q_i. According to the continuity equation, we have

$$Q_i = Q_{i+1} + q_i \tag{5.4}$$

The relationship between flow rate and head loss of each pipe is

$$h_{fi} = \frac{Q_i^2 l_i}{K_i^2}$$

The total head loss of series flow equals to the sum of the head loss in each pipe, that is

$$H = h_f = \sum_{i=1}^{n} h_{fi} = \sum_{i=1}^{n} \frac{Q_i^2 l_i}{K_i^2}$$
$$= \frac{Q_1^2 l_1}{K_1^2} + \frac{Q_2^2 l_2}{K_2^2} + \cdots\cdots + \frac{Q_n^2 l_n}{K_n^2} \tag{5.5}$$

If there is no leakage in each pipe, then

$$H = h_f = Q^2 \sum_{i=1}^{n} \frac{l_i}{K_i^2} \tag{5.6}$$

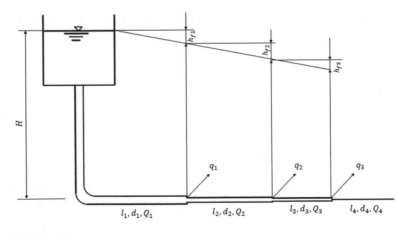

Fig. 5.3 Series flow

Example 5.3 Considering the Example 5.2 with series flow.

Solution

Assuming that the length of the pipe with diameter $d_1 = 350$ mm is l_1, so the length of the pipe with diameter $d_2 = 400$ mm is $l_2 = l - l_1$. Then according to Eq. (5.6), we have

$$H = Q^2 \left(\frac{l_1}{K_1^2} + \frac{l - l_1}{K_2^2} \right)$$

Namely $30 = 250^2 \times \left(\frac{l_1}{1727^2} + \frac{2500 - l_1}{2464^2} \right)$

So

$$l_1 = 400 \text{ m}$$

Therefore, for series flow, the length of the pipe with diameter $d_1 = 350$ mm is 400 m, and the length of the pipe with diameter $d_2 = 400$ mm is $2500 - 400 = 2100$ m.

5.2.2 Parallel Flow

As shown in Fig. 5.4, there are three pipes in parallel between point A and B. The total flow rate is Q. The diameter of each pipe is d_1, d_2 and d_3 respectively, and the length is l_1, l_2 and l_3 respectively. The flow rate of each pipe is Q_1, Q_2 and Q_3 respectively. The head loss is h_{f1}, h_{f2} and h_{f3} respectively. The piezometric head difference between point A and B is h_f. Because there exists only one piezometric head for each point, the piezometric head difference between point A and B through different pipes is invariable. Thus, the characteristic of parallel flow is that the head loss of each parallel pipe is same.

$$h_f = h_{f1} = h_{f2} = h_{f3} \tag{5.7}$$

Fig. 5.4 Parallel flow

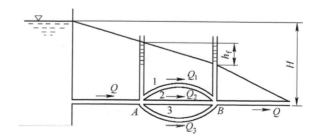

So

$$\frac{Q_1^2 l_1}{K_1^2} = \frac{Q_2^2 l_2}{K_2^2} = \frac{Q_3^2 l_3}{K_3^2} = h_f \tag{5.8}$$

According to continuity equation, we have

$$Q = Q_1 + Q_2 + Q_3 \tag{5.9}$$

Combing Eq. (5.8) with Eq. (5.9), we can solve many problems about the hydraulic calculation of parallel flow.

Example 5.4 As shown in Fig. 5.4, The diameter of each pipe is $d_1 = 150$ mm, $d_2 = 150$ mm and $d_3 = 200$ mm respectively, and the length is $l_1 = 500$ m, $l_2 = 350$ m and $l_3 = 1000$ m respectively. The overall flow rate of this parallel piping system is $Q = 80$ L/s, and each of these pipes belongs to normal pipe. Find the flow rate of each pipe and the head loss of this parallel piping system.

Solution

According to Table 5.2, $K_1 = K_2 = 158.4$, and $K_3 = 341.0$.

Assuming that the flow rate of pipe 1 is Q_1, according to Eq. (5.8), the flow rate of pipe 2 is $Q_2 = Q_1 \frac{K_2}{K_1} \sqrt{\frac{l_1}{l_2}} = Q_1 \times \sqrt{\frac{500}{350}} = 1.195\, Q_1$

The flow rate of pipe 3 is $Q_3 = Q_1 \frac{K_3}{K_1} \sqrt{\frac{l_1}{l_3}} = Q_1 \times \frac{341.0}{158.4} \times \sqrt{\frac{500}{1000}} = 1.522\, Q_1$

The overall flow rate is $Q = Q_1 + Q_2 + Q_3 = Q_1 + 1.195 Q_1 + 1.522 Q_1 = 3.715 Q_1$

So

$$Q_1 = 21.5 \text{ L/s}, \ Q_2 = 25.8 \text{ L/s}, \ Q_3 = 32.7 \text{ L/s}$$

The head loss is $h_f = \frac{Q_1^2 l_1}{K_1^2} = \frac{21.5^2 \times 500}{158.4^2} = 9.2$ m H_2O

5.2.3 Uniformly Variable Mass Outflow

As shown in Fig. 5.5, assuming that the outlet flow rate is Q_T, and the total leakage flow rate is Q_P. If the leakage flow rate per unit length along the pipe is same, $\frac{Q_P}{l} = q$ is constant. This kind of flow is defined as uniformly variable mass outflow.

Select arbitrary point M for analysis, and the distance from point M to point A is x. The flow rate in point M is

$$Q_M = Q_T + Q_P - \frac{Q_P}{l} x$$

Fig. 5.5 Uniformly variable
mass outflow [2]

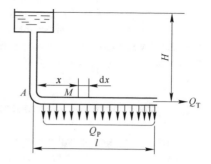

Then

$$dh_f = \frac{Q_M^2 dx}{K^2} = \frac{1}{K^2}\left(Q_T + Q_P - \frac{Q_P}{l}x\right)^2 dx$$

Integrate above equation, we have

$$h_f = \frac{1}{K^2}\int_0^l \left(Q_T + Q_P - \frac{Q_P}{l}x\right)^2 dx \qquad (5.10)$$

$$= \frac{l}{K^2}\left[Q_T^2 + Q_T Q_P + \frac{1}{3}Q_P^2\right]$$

It can approximately equal to

$$h_f = \frac{l}{K^2}(Q_T + 0.55Q_P)^2 \qquad (5.11)$$

Combining the form of Chezy formula, we rewrite the equation as follows:

$$h_f = \frac{Q_c^2 l}{K^2} \qquad (5.12)$$

where $Q_c = Q_T + 0.55Q_P$.

When $Q_T = 0$, Eq. (5.10) can be simplified as

$$h_f = \frac{1}{3}\frac{Q_P^2 l}{K^2} \qquad (5.13)$$

According to Eq. (5.13), the head loss of uniformly variable mass outflow is just a third of the head loss of invariable mass flow with same total flow rate. The reason is that the volumetric flow rate of variable mass outflow decreases gradually along the pipe.

5.2.4 Pipe Network Geometry

The pipe networks for water supply have mainly the following two types of configurations: branched or tree-like configuration (a) and looped configuration (b), as shown in Fig. 5.6.

A branched network, or a tree network, is a distribution system having no loops. Such a network is commonly utilized for rural water supply. The simplest branched network is a radial network consisting of several distribution mains emerging out from a common input point.

A pipe network in which there are one or more closed loops is called a looped network. Looped networks are preferred from the reliability point of view. If one or more pipelines are closed for repair, water can still reach the consumer by a circuitous route incurring more head loss. This feature is absent in a branched network. With the changing demand pattern, not only the magnitudes of the discharge but also the flow directions change in many links.

5.3 Orifice Flow

Orifice flow refers to fluid outflow from an orifice [3, 4]. When fluid flows out through an orifice into atmosphere, we call this kind of flow as free discharge. When fluid flows out through an orifice into another liquid, we call this kind of flow as submerged discharge.

According to the diameter, orifice can be divided into small orifice and large orifice. Assuming that the head acting on the orifice cross section is H, orifice diameter is d. When $d < \frac{H}{10}$, the orifice belongs to small orifice; when $d > \frac{H}{10}$, the orifice belongs to large orifice.

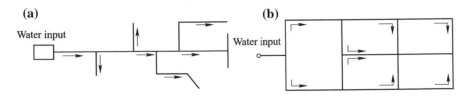

Fig. 5.6 Pipe networks

If fluid discharge through an orifice on a container wall might not be affected by sharp edge it has and the thickness of container wall is smaller than $3d$, this kind of flow is still supposed to thin-plate orifice flow. Otherwise the discharge of orifice on thick wall of which the thickness is usually bigger than $3d$ is supposed to nozzle flow, which will be discussed in the next section.

5.3.1 Steady Discharge Through Thin-Plate Orifice

5.3.1.1 Free Discharge of Thin-Plate Orifice
The free discharge through thin-plate orifice is shown in Fig. 5.7. Assuming that the head H is constant, the diameter of the orifice is d and the area is A.

When the fluid enters the orifice, the streamlines cannot change direction sharply due to the inertial effect and bend gradually along a smooth curvet to converge towards the orifice axis until to a cross section c–c with a roughly distance of $0.5d$ from inlet.

The cross section c–c is called the vena contracta. Assuming that the area of c–c is A_c, then

$$\frac{A_c}{A} = \varepsilon < 1, \tag{5.14}$$

where ε is contraction coefficient.

Consider the flow between section O–O and c–c in Fig. 5.7. Select a datum reference O'–O' on the centerline of the orifice, then we write Bernoulli equation between the two sections as

$$H + \frac{p_a}{\gamma} + \frac{\alpha_0 v_0^2}{2g} = 0 + \frac{p_c}{\gamma} + \frac{\alpha_c v_c^2}{2g} + h_l \tag{5.15}$$

Fig. 5.7 Free discharge of thin-plate orifice

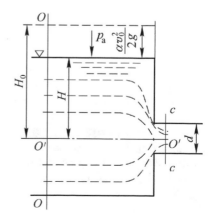

The liquid on cross section c–c is connected with the atmosphere, so we assume $P_c = P_a$. Since friction loss is very small, only the minor head loss is taken into consideration, that is, $h_l = h_r = \zeta \frac{v_c^2}{2g}$, in which ζ is minor loss coefficient of orifice. We assume $\alpha_0 = \alpha_c = 1.0$, because of the uniform velocity distribution on the orifice. Therefore, the above equation may be transformed into

$$H + \frac{v_0^2}{2g} = \frac{v_c^2}{2g} + \zeta \frac{v_c^2}{2g} = (1 + \zeta) \frac{v_c^2}{2g}$$

Then

$$v_c = \frac{1}{\sqrt{1+\zeta}} \sqrt{2g \left(H + \frac{v_0^2}{2g} \right)} \qquad (5.16)$$

We express the above equation as follows:

$$v_c = \varphi \sqrt{2gH_0}, \qquad (5.17)$$

where $\varphi = \frac{1}{\sqrt{1+\zeta}}$ is discharge coefficient for velocity, and $H_0 = H + \frac{v_0^2}{2g}$ is effective head.

Compared with v_c, v_0 is much smaller and always neglected, so v_c can be approximately expressed as follows.

$$v_c = \varphi \sqrt{2gH} \qquad (5.18)$$

Thus, according to Eqs. (5.17) and (5.18), the flow rate of orifice flow is

$$Q = \varepsilon A \varphi \sqrt{2gH_0} = \mu A \sqrt{2gH_0} \qquad (5.19)$$

Or

$$Q = \varepsilon A \varphi \sqrt{2gH} = \mu A \sqrt{2gH}, \qquad (5.20)$$

where $\mu = \varepsilon \varphi$ is discharge coefficient for flow rate.

With different types of orifice and resistance, there will be different values of discharge coefficient. According to many experimental results, the discharge coefficients take values from $\varphi = 0.97 \sim 0.98$, $\mu = 0.58 \sim 0.62$.

5.3.1.2 Submerged Discharge

As shown in Fig. 5.8, for submerged discharge, the water jet passing through the orifice will expand rapidly and flow into another liquid.

With 0–0 as the datum reference, the Bernoulli equation between section 1–1 and 2–2 is

Fig. 5.8 Submerged discharge

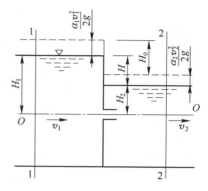

$$H_1 + \frac{p_1}{\gamma} + \frac{\alpha_1 v_1^2}{2g} = H_2 + \frac{p_2}{\gamma} + \frac{\alpha_2 v_2^2}{2g} + h_l, \tag{5.21}$$

where, $h_l = h_r = \zeta_s \frac{v_c^2}{2g}$. ζ_s is minor loss coefficient of submerged discharge, and it includes two parts: minor loss coefficient of jet contraction ζ and minor loss coefficient of jet sudden expansion ζ_E. We take $\zeta_E = 1$, so $\zeta_s = \zeta + 1$. We assume $v_1 = v_2$, $\alpha_1 = \alpha_2 = 1.0$, $p_1 = p_2 = p_a$. Then, Eq. (5.21) can be transformed into

$$H_1 = H_2 + (1 + \zeta)\frac{v_c^2}{2g}$$

then

$$v_c = \frac{1}{\sqrt{1+\zeta}}\sqrt{2g(H_1 - H_2)} = \varphi\sqrt{2gH}, \tag{5.22}$$

where $H = H_1 - H_2$, which is elevation head difference between upstream and downstream.

Thus, the flow rate of submerged discharge is

$$Q = A_c v_c = \varepsilon A \varphi \sqrt{2gH} = \mu A \sqrt{2gH},$$

where μ is discharge coefficient of submerged discharge for flow rate.

Considering that the head of downstream has little influence on the jet contraction and minor loss, we assume the ε, φ, μ of submerged discharge are basically the same as free discharge.

Example 5.5 Assuming that there is free discharge of a thin-plate orifice with diameter $d = 50$ mm, head $H = 1$ m. Find the flow rate of free discharge. If it is submerged discharge, and the head after the orifice discharge is $H_2 = 0.4$ m, find the corresponding flow rate.

Solution

Neglecting the velocity head v_0, and taking discharge coefficient for flow rate $\mu = 0.62$, according to Eq. (5.20), we have

$$Q = \mu A \sqrt{2gH} = \mu \times \frac{\pi}{4} d^2 \sqrt{2gH} = 0.62 \times \frac{\pi}{4} \times 0.05^2 \times \sqrt{2 \times 9.8 \times 1}$$
$$= 0.0054 \text{ m}^3/\text{s} = 5.4 \text{ L/s}$$

When it comes to the submerged discharge, the head difference between upstream and downstream is $Z = H - H_2 = 1 - 0.4 = 0.6$ m, then the flow rate is

$$Q = \mu A \sqrt{2gZ} = \mu \times \frac{\pi}{4} d^2 \sqrt{2gZ} = 0.62 \times \frac{\pi}{4} \times 0.05^2 \times \sqrt{2 \times 9.8 \times 0.6}$$
$$= 0.00417 \text{ m}^3/\text{s} = 4.17 \text{ L/s}$$

5.3.2 Discharge Through Big Orifice

As shown in Fig. 5.9, the head in each point on vena contracta is different for big orifice free discharge. In most cases of engineering applications, for big orifice flow rate calculation, we utilize the formula which is similar to small orifice. However, the discharge coefficient for flow rate is determined by the experiments. Thus, the equation to calculate the flow rate is expressed as follows:

$$Q = \mu' A \sqrt{2gH}, \tag{5.23}$$

where μ' is discharge coefficient of big orifice for flow rate ranging from $\mu' = 0.6 \sim 0.9$.

Fig. 5.9 Discharge through big orifice

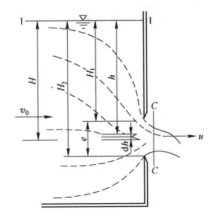

5.3.3 Unsteady Discharge Through Orifice

Unsteady discharge means that the free surface of the container lowers gradually. We usually pay more attention to the discharge time for unsteady discharge calculation.

As shown in Fig. 5.10, we assume the area of the container cross section is A, the area of orifice is a, the height from free surface of container to orifice is y. During the time dt, the flow rate of orifice discharge can be calculated by Eq. (5.20) shown as follows:

$$Q = \mu a \sqrt{2gy}$$

Assume that the free surface decreases dy during the time dt. According to the continuity equation, the outflow volume from orifice equals the decreasing volume in the container. We have

$$A dy = Q dt = \mu a \sqrt{2gy} dt$$

Then

$$dt = \frac{A dy}{\mu a \sqrt{2gy}}$$

The time t of the free surface decreasing from H_1 to H_2 is indicated after integrating the above equation as

Fig. 5.10 Unsteady discharge through orifice

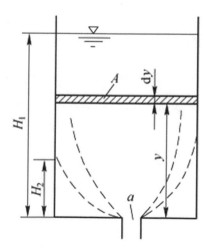

$$t = \int_{H_2}^{H_1} \mathrm{d}t = \int_{H_2}^{H_1} \frac{A}{\mu a \sqrt{2g}} \frac{\mathrm{d}y}{\sqrt{y}}$$

$$= \frac{A}{\mu a \sqrt{2g}} [2\sqrt{y}]_{H_2}^{H_1} = \frac{2A}{\mu a \sqrt{2g}} \left[\sqrt{H_1} - \sqrt{H_2} \right] \tag{5.24}$$

When $H_2 = 0$, the time of emptying the water in the container is

$$t = \frac{2A\sqrt{H_1}}{\mu a \sqrt{2g}} = \frac{2AH_1}{\mu a \sqrt{2gH_1}} = \frac{2V}{Q_{\max}}, \tag{5.25}$$

where V is the emptying volume of the container, and $Q_{\max}$ is the maximum flow rate at starting time.

Equation (5.25) shows that the emptying time of unsteady discharge is twice the time for same outflow volume in steady discharge with the head H_1.

5.4 Nozzle Flow

When a short pipe with length $l = (3 \sim 4)d$ (where d is the pipe diameter) and same cross section with orifice is connected to the orifice of a container, the outflow in this case is defined as nozzle discharge. The nozzle can be classified into cylindrical outer nozzle (Fig. 5.11a), cylindrical inner nozzle (Fig. 5.11b), conical contracted nozzle (Fig. 5.11c), conical expanding nozzle (Fig. 5.11d), and streamline shape nozzle (Fig. 5.11e).

5.4.1 Steady Discharge Through Cylindrical Outer Nozzle

A cylindrical outer nozzle is shown in Fig. 5.12. The water jet passing through the nozzle will contract at first to generate a vena contracta c–c, the distance from inlet is about $L_c = 0.8d$. The vacuum phenomenon occurs as well. Then the jet can fill

Fig. 5.11 Different types of nozzle flow

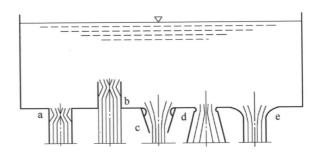

Fig. 5.12 Steady discharge through cylindrical outer nozzle

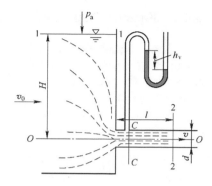

the nozzle gradually. So only the minor loss would be involved in hydraulic calculation.

Assuming that the area of nozzle cross section is A. With 0–0 as the datum reference, the Bernoulli equation between section 1–1 and 2–2 is

$$H + \frac{p_a}{\gamma} + \frac{\alpha_1 v_1^2}{2g} = \frac{p_a}{\gamma} + \frac{\alpha v^2}{2g} + h_l, \qquad (5.26)$$

where $h_l = h_r = \sum \zeta \frac{v^2}{2g}$, $\sum \zeta$ is the total minor loss coefficient, including the contraction and the expansion of the jet. We take $\alpha_1 = \alpha = 1.0$ and replace v_1 with v_0. Assuming that $H_0 = H + \frac{v_0^2}{2g}$, we have

$$H_0 = \left(1 + \sum \zeta\right) \frac{v^2}{2g}$$

The velocity can be expressed as

$$v = \frac{1}{\sqrt{1 + \sum \zeta}} \sqrt{2gH_0} = \varphi \sqrt{2gH_0} \qquad (5.27)$$

Compared with v, v_0 is much smaller and can be neglected, so we have

$$v = \varphi \sqrt{2gH} \qquad (5.28)$$

The flow rate of nozzle discharge is

$$Q = Av = \varphi A \sqrt{2gH} = \mu A \sqrt{2gH}, \qquad (5.29)$$

where φ is discharge coefficient for velocity of cylindrical outer nozzle, μ is discharge coefficient for flow rate of cylindrical outer nozzle, and $\mu - \varphi$. According to the tests, the value of discharge coefficient for flow rate is $\mu = 0.82$.

5.4.2 Vacuum Degree of Nozzle

The discharge coefficient for flow rate of orifice is smaller than the one of nozzle due to vacuum generation in the nozzle. The flow rate can be improved a lot after connecting a nozzle on thin-plate orifice.

As shown in Fig. 5.12, when U-tube manometer is connected with the vena contracta c–c, the height of the liquid column in U-tube h_v equals $0.75H_0$ according to experimental results. The value can also be analyzed by theoretical analysis as follows:

As shown in Fig. 5.12, with 0–0 as the datum reference, the Bernoulli equation between section 1–1 and c–c is

$$H + \frac{p_a}{\gamma} + \frac{\alpha_1 v_1^2}{2g} = \frac{p_c}{\gamma} + \frac{\alpha_c v_c^2}{2g} + \zeta \frac{v_c^2}{2g}$$

Neglecting $\frac{\alpha_1 v_1^2}{2g}$ and assuming that $\alpha_c = 1.0$, we have

$$\frac{p_a - p_c}{\gamma} = (1 + \zeta)\frac{v_c^2}{2g} - H \tag{5.30}$$

According to the continuity equation, we have

$$v_c^2 = \left(\frac{A}{A_c}\right)^2 v^2 = \frac{v^2}{\varepsilon^2}$$

According to Eq. (5.28), we have

$$\frac{v^2}{2g} = \varphi^2 H$$

Equation (5.30) can be expressed as follows:

$$\frac{p_a - p_c}{\gamma} = (1 + \zeta)\frac{\varphi^2 H}{\varepsilon^2} - H$$

Assuming that $\zeta = 0.06$ (minor loss coefficient of gradual contraction short pipe in Table 4.10), $\varepsilon = 0.64$ and $\varphi = 0.82$ usually. Then we have

$$\frac{p_a - p_c}{\gamma} = 0.74H \approx 0.75H_0 \tag{5.31}$$

It can be seen that the bigger H_0 is, the higher vacuum will be. The vacuum degree on the vena contracta is 75% of the acting head, which indicates that the function of nozzle is equivalent to increasing the acting head of orifice free discharge by 75%. Therefore, the flow rate of nozzle discharge is much bigger than that of corresponding orifice.

According to the above analysis, the higher vacuum can lead to bigger flow rate. However, in order to ensure the nozzle working normally, the highest vacuum in the nozzle should be restricted. According to the experimental results, the vacuum on the vena contracta should not exceed 7 m H_2O, that is

$$\frac{p_a - p_c}{\gamma} = 0.75H_0 \leq 7, \ H_0 \leq 9 \text{ m}$$

Second, there is restriction on the length of nozzle as well. The most suitable length l should be about $3 \sim 4d$.

Example 5.6 As shown in Fig. 5.13, there is a water tank with three cylindrical drainage orifices. The diameter of the drainage orifice is $d = 0.2$ m. The thickness of the wall is $l = 0.7$ m. The head above the centerline of orifice is $H = 1.5$ m. Neglecting the moving velocity v_0, find the flow rate of the drainage orifices.
Solution

Since the thickness of the wall $l = 3.5\,d$, the drainage of this water tank can be regarded as nozzle discharge. Taking discharge coefficient for flow rate $\mu = 0.82$, the flow rate of each drainage orifice is

$$q = \mu A \sqrt{2gH} = 0.82 \times \frac{\pi}{4} \times 0.2^2 \times \sqrt{2 \times 9.8 \times 1.5} = 0.14 \text{ m}^3/\text{s}$$

The overall flow rate of these three drainage orifices are

$$Q = 3q = 0.42 \text{ m}^3/\text{s}$$

Since the head above the centerline of orifice is $H = 1.5$ m < 9 m, the discharge of drainage orifice functions normally.

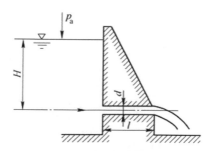

Fig. 5.13 Drainage of a water tank

5.4.3 Other Types of Nozzle Discharge

For other types of nozzle discharge, their velocity, and flow rate formulas are the same with that of the cylindrical nozzle. The difference is the discharge coefficient, below are several commonly used nozzles.

5.4.3.1 Streamline Shape Nozzle
As shown in Fig. 5.14a, the discharge coefficient for velocity is $\varphi = \mu = 0.97$, and it is suitable for the case of small head loss, big flow rate, and uniform velocity distribution at the outlet.

5.4.3.2 Conical Expanding Nozzle
As shown in Fig. 5.14b, when $\theta = 5° \sim 7°$, $\varphi = \mu = 0.42 \sim 0.50$. It is suitable for the case of transforming part of kinetic energy to pressure energy, such as the diffuser pipe of ejector.

5.4.3.3 Conical Contracted Nozzle
As shown in Fig. 5.14c, the discharge is dependent on the contraction angle. When $\theta = 30°24'$, $\varphi = 0.963$, $\mu = 0.943$ is the maximum magnitude. It is suitable for the case of speeding up the ejecting velocity, such as firefighting water gun.

5.5 Problems

5.1 As shown in Fig. 5.15, there is a cast iron pipe ($\Delta = 0.4$ mm) with constant diameter $d = 500$ mm, and the length $l = 100$ m. Water flow belongs to turbulent flow in rough pipes (region V). The minor loss coefficients of inlet and outlet are 0.5 and 1.0 respectively, and for each elbow $\zeta = 0.3$. The height from the upstream to the downstream is $H = 5$ m. Find the flow rate Q of the pipe.

Fig. 5.14 Common nozzle discharge

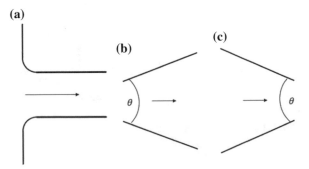

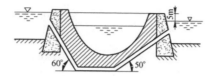

Fig. 5.15 Problem 5.1

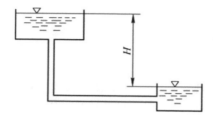

Fig. 5.16 Problem 5.2

5.2 The water flows from the high-level reservoir to the low-level reservoir, as shown in Fig. 5.16. Assuming that $H = 12$ m. There is a clean steel pipe with the length $l = 300$ m and the diameter $d = 100$ mm. Find the flow rate of the pipe. If the flow rate $Q = 150$ m^3/h, what is the diameter of the pipe?

5.3 As shown in Fig. 5.17, assuming that the total head of this piping system is $H = 12$ m, the diameter and length of each pipe is $d_1 = 250$ mm, $l_1 = 1000$ m, $d_2 = 200$ mm, $l_2 = 650$ m, $d_3 = 150$ mm, $l_3 = 750$ m respectively. Find the head loss in each pipe, and draw piezometric head line. The pipe is clean with minor head loss neglected.

5.4 As shown in Fig. 5.18, there is a parallel piping system. The flow rate of each pipe is $Q_1 = 50$ L/s, $Q_2 = 30$ L/s respectively with the length $l_1 = 1000$ m,

Fig. 5.17 Problem 5.3

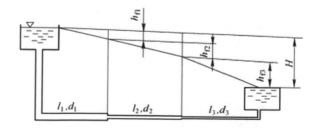

Fig. 5.18 Problem 5.4

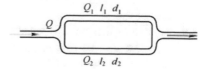

Fig. 5.19 Problem 5.5

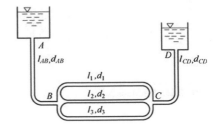

Fig. 5.20 Problem 5.6

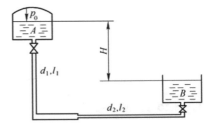

and $l_2 = 500$ m. The diameter is $d_1 = 200$ mm. The pipe is clean. Find the diameter d_2.

5.5 As shown in Fig. 5.19, there are three different pipes between B and C. The dimension of each pipe is $d_1 = 300$ mm, $l_1 = 500$ m, $d_2 = 250$ mm, $l_2 = 300$ m, $d_3 = 400$ mm, $l_3 = 800$ m, $d_{AB} = 500$ mm, $l_{AB} = 800$ m, $d_{CD} = 500$ mm, $l_{CD} = 400$ m. The pipes all belong to normal pipes. The flow rate in point B is 250 L/s. Find the total head loss in this piping system.

5.6 The relative pressure on the surface of water tank A is $p_0 = 1.313 \times 10^5$ Pa, and water flows from A to B through a series of different pipes, as shown in Fig. 5.20. Assuming that $H = 8$ m. The dimension of the pipes are $d_1 = 200$ mm, $l_1 = 200$ m, $d_2 = 100$ mm, $l_2 = 500$ m respectively. The pipes all belong to normal pipes. Only the minor head loss from the valve is taken into consideration. Find the flow rate Q of the pipe.

5.7 The pumping station utilizes a pipe with the diameter 60 cm to transport the water, and the friction loss is 27 m. To reduce the loss, another pipe with same length is connected in parallel with the first pipe, and correspondingly the total head loss decreases to 9.6 m. Assuming that both pipes have same friction factor and in either case, the total flow rate do not change. Find the diameter of second pipe.

5.8 As shown in Fig. 5.21, water flows from one reservoir to another. Assuming that $H = 24$ m, $l_1 = l_2 = l_3 = l_4 = 100$ m, $d_1 = d_2 = d_4 = 100$ mm, $d_3 = 200$ mm. The friction factor $\lambda_1 = \lambda_2 = \lambda_4 = 0.025$, $\lambda_3 = 0.02$. Except the minor head loss from valve, others can be neglected. Find: (1) the flow rate of the pipe when the minor loss coefficient of the valve is $\zeta = 30$. (2) the flow rate of the pipe when the valve is closed.

5.9 Assuming that there is steady discharge through a thin-plate orifice with the diameter $d = 10$ mm in the tank. The flow rate is $Q = 200$ cm^3/s. Find the height H of the water in the tank.

Fig. 5.21 Problem 5.8

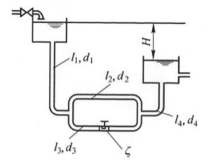

Fig. 5.22 Problem 5.10

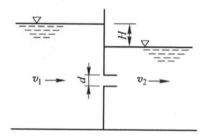

Fig. 5.23 Problem 5.11

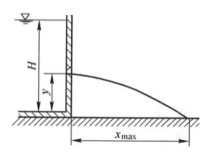

5.10 As shown in Fig. 5.22, a reservoir is divided into two parts. Assuming that the diameter of small orifice is $d = 200$ mm, and $v_1 \approx v_2 \approx 0$. The water level difference from the upstream to the downstream is $H = 2.5$ m. Find the flow rate of the small orifice.

5.11 As shown in Fig. 5.23, there is a water tank with the height from the free surface to the bottom H, where the orifice on the side wall should be fixed in order to cover a maximum horizontal range for the jet. Find x_{max}.

5.12 Assuming that the diameter of the orifice $d = 100$ mm, head $H = 3$ m. The velocity on vena contracta is $v_c = 7$ m/s, and the flow rate is $Q = 36$ L/s. Find: (1) the discharge coefficient for velocity φ and contraction coefficient ε of the orifice. (2) the flow rate if a cylindrical outer nozzle is connected to the orifice, and the discharge coefficient for flow rate is $\mu = 0.82$.

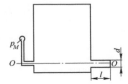

Fig. 5.24 Problem 5.13

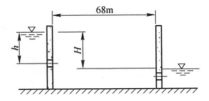

Fig. 5.25 Problem 5.15

5.13 There is an enclosed container with liquid inside (the specific weight $\gamma = 7850$ N/m^3). A cylindrical outer nozzle with the length $l = 100$ mm, the diameter $d = 30$ mm is fixed on the section O–O, as shown in Fig. 5.24. The manometer is fixed 0.5 m higher than the section O–O with $p_M = 4.9 \times 10^4$ Pa. Find the velocity and the flow rate at starting time.

5.14 There is a rectangular tank with length $l = 3$ m, width $B = 2$ m, and depth of the water $H = 1.5$ m. Two orifices with the diameter $d = 100$ mm are fixed at the bottom of the tank. Find the time when the free surface decreases 1 m.

5.15 As shown in Fig. 5.25, find the filling or emptying time of a chamber in ship lock. Assuming that the length and the width of the chamber are 68 m, 12 m respectively. The area of the upstream inlet orifice is 3.2 m^2, and the height from the surface of the upstream to the centerline of the orifice is $h = 4$ m. The water level of the upstream and the downstream remain unchanged and their difference is $H = 7.0$ m.

References

1. Song, H.: Engineering Fluid Mechanics and Environmental Application. Metallurgical Industry Press, Beijing (2016)
2. Xie, Z.: Engineering Fluid Mechanics, 4th edn. Metallurgical Industry Press, Beijing (2014)
3. White, F.M.: Fluid Mechanics, 7th edn. McGraw-Hill, New York (2011)
4. Fox, R.W., McDonald, A.T., Pritchard, P.J.: Introduction to Fluid Mechanics, vol. 5, 8th edn. Wiley, New York (2011)

Fundamentals of Fluid Mechanics Through Porous Media

6

Abstract

Fluid flow through porous media widely exists in nature and artificial materials, and its theory has been used in all sorts of scientific and technological fields, such as soil mechanics, petroleum engineering, mineral engineering, environmental engineering, geothermal engineering, water supply engineering, chemical industry, micro machine, and so on. In this chapter, we will first give some basic concepts of fluid flow through porous media, such as porosity and compressibility of porous media. Then we will introduce Darcy's law and mathematical model of fluid flow through porous media. Finally, we will utilize some principles of seepage mechanics to discuss solutions for planar one-directional flow and planar radial flow.

Keywords

Porous media · Porosity · Compressibility · Darcy's law · Hydraulic conductivity

6.1 Several Concepts

Fluid flow through porous media is an important branch of fluid mechanics, which is based on the combination of porous media theory, surface physics, physical chemistry, and biology.

The characteristics of fluid flow through porous media are as follows. First, due to the large specific surface area and strong surface effect in porous media, viscous effect should be taken into consideration at all times. Second, the fluid compressibility should also be considered due to the high pressure in the formation. In addition, molecular force cannot be neglected sometimes due to the complex pore shape, big resistance and capillary force. Finally, there often exist complex physical and chemical processes while fluid flowing through porous media.

6.1.1 Porous Media

Porous media refer to solids which contain various types of capillary systems such as void, fracture, and so on [1, 2]. To sum up, the following points are often used to describe porous media:

(1) Porous medium is the space occupied by multiphase medium, where the solid phase is called solid skeleton, the unoccupied space is called pore which can be filled with gas, liquid or multiphase fluid.

(2) The solid phase and pores should extend throughout the whole medium. This means that taking a proper size volume from the medium, then this volume must contain a certain proportion of solid particles and pores.

(3) The pore space should be interconnected so that the fluid can flows in them, and this kind of pore space is called effective pore space, while disconnected pore space or dead pore is called void pore space. For flow in porous media, void pore space is actually regarded as solid skeleton.

The common pore structure is intergranular pore structure, which consists of solid particles with different sizes and shapes. The spaces between particles are usually occupied by cement; the unoccupied parts are the pores, which are the reservoirs and flow channels for fluids, as shown in Fig. 6.1.

6.1.2 Porosity

Pore structure in the porous medium provides space for fluids to store and flow in, and porosity is a measure of the void space in the porous medium. The total volume V_b consists of pore volume V_p and solid particle volume V_s,

Fig. 6.1 Intergranular pore structures

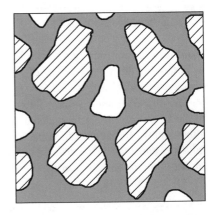

$$V_b = V_p + V_s \tag{6.1}$$

Porosity ϕ refers to the ratio of the pore volume V_p to the total volume V_b

$$\phi = \frac{V_p}{V_b} \times 100\% \tag{6.2}$$

$$\phi = \frac{V_b - V_s}{V_b} \times 100\% = \left(1 - \frac{V_s}{V_b}\right) \times 100\% \tag{6.3}$$

6.1.3 Compressibility of Porous Media

Strictly speaking, all material have elasticity and can be compressed, porous media are no exception. For example, rocks are porous media, when under external pressure they will deform and become more compacted, thus the pore volume becomes smaller.

The elastic compression coefficient C_f is usually used to describe the elastic state

$$C_f = \frac{\frac{dv_p}{V_p}}{dp}, \tag{6.4}$$

where V_p is the pore volume and

$$\frac{dV_p}{V_p} = \frac{d\phi}{\phi} \tag{6.5}$$

Thus

$$C_f = \frac{1}{\phi}\frac{d\phi}{dp}, \tag{6.6}$$

where C_f is the rock compression coefficient, generally ranging $(1 \sim 2) \times 10^{-4}$ MPa^{-1}; V_b is the total volume of the rock.

6.2 Darcy's Law

When fluid flows in the porous media, there must be energy loss due to viscous effect. Darcy proposed a basic relationship between energy loss and velocity based on the experimental results from 1852 to 1855 [3, 4]. This relationship is called Darcy's law and is the most basic relation for fluid flow through porous media.

6.2.1   Description of Darcy's Law and Permeability

The device of Darcy experiment is shown in Fig. 6.2. The main body of the device is a vertical cylinder with an opened upper end, and two piezometer tubes are connected to its side wall. A filter plate C is fixed at a certain distance from the bottom, on which there are homogeneous sands. Water flows in from the upper end, then flows through the sands and flows out from short tube T into vessel V which is used to measure the flow rate. The overflow pipe B is used to keep the water level constant in the cylinder.

After a certain period of time, the flowrates of the upper end the tube T becomes identical and the water level in piezometer is constant, which means that the flow is steady. Since the velocity is quite small, so velocity head loss can be ignored. Therefore, total head equals the head of the piezometer, and head loss h_w equals the head difference of the two piezometers,

$$h_w = h_1 - h_2 \tag{6.7}$$

$$h_1 = z_1 + \frac{p_1}{\rho g} + \frac{v_1^2}{2g} \tag{6.8}$$

$$h_2 = z_2 + \frac{p_2}{\rho g} + \frac{v_2^2}{2g} \tag{6.9}$$

Hydraulic gradient J can be expressed using hydraulic gradient of the piezometer

Fig. 6.2 Darcy experiment device [2]

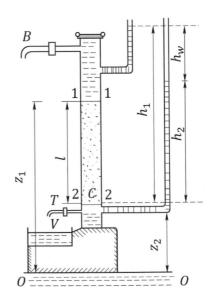

$$J = \frac{h_w}{l} = \frac{h_1 - h_2}{l} = \frac{(z_1 - z_2) + \frac{p_1 - p_2}{\rho g}}{l}$$

$$= \frac{l + \frac{p_1 - p_2}{\rho g}}{l} = 1 +, \frac{p_1 - p_2}{l \rho g} \tag{6.10}$$

where l is the distance between section 1-1 and 2-2; h_w is the head loss between the two above sections; h_1 and h_2 is the piezometer head of section 1-1 and 2-2, respectively.

Based on the analysis of a large amount of experiment data, Darcy thought that the flow rate Q was directly proportional to section area A and hydraulic gradient J, and was relevant to soil properties, thus he proposed a formula as follow:

$$Q = kAJ, \tag{6.11}$$

where k is the hydraulic conductivity, which reflects the property of porous medium and has the dimension of velocity.

The average velocity of the section is

$$v = \frac{Q}{A} = kJ = k\left(1 + \frac{p_1 - p_2}{\rho g l}\right) = k\left(\frac{\rho g + (p_1 - p_2)/l}{\rho g}\right) \tag{6.12}$$

Experiments show that hydraulic conductivity k is directly proportional to fluid specific weight ρg and inversely proportional to fluid viscosity μ, using K as the ratio coefficient, we have

$$k = \frac{K \rho g}{\mu}, \tag{6.13}$$

where K is called permeability and only relevant to the structure of the porous medium [5], its dimension is $[L]^2$.

Substituting Eq. (6.13) into Eq. (6.12), Darcy's law can be obtained

$$v = \frac{K}{\mu}\left(\frac{p_1 - p_2}{l} + \rho g\right) = \frac{K}{\mu}\left(\frac{\partial p}{\partial z} + \rho g\right) \tag{6.14}$$

If the sand layer is horizontal, then gravity can be ignored and Darcy's law can be simplified as

$$Q = Av = A\frac{K}{\mu}\frac{p_1 - p_2}{l} = A\frac{K}{\mu}\frac{dp}{dx} \tag{6.15}$$

So flow rate Q and pressure difference have a linear relationship, and Darcy's law is linear. If pressure difference and section area are known, then there are only two factors influencing the resistance: one is fluid viscosity, the other is porous medium property, namely permeability.

6.2.2   Deduction from Navier–Stokes Equation

Darcy's law can also be deduced from Navier–Stokes equation. Since the velocities v_x, v_y v_z are not real velocities, so they must be replaced with $\frac{v_x}{\phi}$, $\frac{v_y}{\phi}$ and $\frac{v_z}{\phi}$ when substituted into N-S equation

$$\frac{1}{\phi}\frac{\partial v_x}{\partial t} + \frac{v_x}{\phi^2}\frac{\partial v_x}{\partial x} + \frac{v_y}{\phi^2}\frac{\partial v_x}{\partial y} + \frac{v_z}{\phi^2}\frac{\partial v_x}{\partial z} = -\frac{1}{\rho}\frac{\partial p}{\partial x} + \frac{\mu}{\rho\phi}\nabla^2 v_x$$

$$\frac{1}{\phi}\frac{\partial v_y}{\partial t} + \frac{v_y}{\phi^2}\frac{\partial v_y}{\partial y} + \frac{v_x}{\phi^2}\frac{\partial v_y}{\partial x} + \frac{v_z}{\phi^2}\frac{\partial v_y}{\partial z} = -\frac{1}{\rho}\frac{\partial p}{\partial y} + \frac{\mu}{\rho\phi}\nabla^2 v_y$$

$$\frac{1}{\phi}\frac{\partial v_z}{\partial t} + \frac{v_z}{\phi^2}\frac{\partial v_z}{\partial z} + \frac{v_y}{\phi^2}\frac{\partial v_z}{\partial y} + \frac{v_x}{\phi^2}\frac{\partial v_z}{\partial x} = -\frac{1}{\rho}\frac{\partial p}{\partial z} + \frac{\mu}{\rho\phi}\nabla^2 v_z - g$$

Define the following statistical averages:

$$\nabla^2 v_x = \left(\frac{1}{c}\right)\left(\frac{v_x}{d^2}\right), \quad \nabla^2 v_y = \left(\frac{1}{c}\right)\left(\frac{v_y}{d^2}\right), \quad \nabla^2 v_z = \left(\frac{1}{c}\right)\left(\frac{v_z}{d^2}\right),$$

where d is average pore radius, and c is dimensionless shape parameter. Substituting the above equation into N-S equation, we have

$$-\frac{1}{\phi}\frac{\partial}{\partial x}\left(\frac{\partial \Phi}{\partial t}\right) + \frac{1}{\phi^2}\frac{\partial}{\partial x}\left[\frac{1}{2}\left(\frac{\partial \Phi}{\partial x}\right)^2 + \frac{1}{2}\left(\frac{\partial \Phi}{\partial y}\right)^2 + \frac{1}{2}\left(\frac{\partial \Phi}{\partial z}\right)^2\right] = -\frac{1}{\rho}\frac{\partial p}{\partial x} + \frac{\mu}{c\rho d^2\phi}\frac{\partial \Phi}{\partial x}$$

$$-\frac{1}{\phi}\frac{\partial}{\partial y}\left(\frac{\partial \Phi}{\partial t}\right) + \frac{1}{\phi^2}\frac{\partial}{\partial y}\left[\frac{1}{2}\left(\frac{\partial \Phi}{\partial x}\right)^2 + \frac{1}{2}\left(\frac{\partial \Phi}{\partial y}\right)^2 + \frac{1}{2}\left(\frac{\partial \Phi}{\partial z}\right)^2\right] = -\frac{1}{\rho}\frac{\partial p}{\partial y} + \frac{\mu}{c\rho d^2\phi}\frac{\partial \Phi}{\partial y}$$

$$-\frac{1}{\phi}\frac{\partial}{\partial z}\left(\frac{\partial \Phi}{\partial t}\right) + \frac{1}{\phi^2}\frac{\partial}{\partial z}\left[\frac{1}{2}\left(\frac{\partial \Phi}{\partial x}\right)^2 + \frac{1}{2}\left(\frac{\partial \Phi}{\partial y}\right)^2 + \frac{1}{2}\left(\frac{\partial \Phi}{\partial z}\right)^2\right] = -\frac{1}{\rho}\frac{\partial p}{\partial z} + \frac{\mu}{c\rho d^2\phi}\frac{\partial \Phi}{\partial z} - g$$

Assuming μ and ρ are constant and integrating the above equation, we have

$$-\frac{1}{\phi}\frac{\partial \Phi}{\partial t} + \phi^2\frac{1}{2}\left[\left(\frac{\partial \Phi}{\partial x}\right)^2 + \left(\frac{\partial \Phi}{\partial y}\right)^2 + \left(\frac{\partial \Phi}{\partial z}\right)^2\right] + \frac{p}{\rho} - \frac{\mu\Phi}{c\rho d^2\phi} + gz = F(t)$$

For steady flow without inertial force, the above equation can be written as

$$\frac{p}{\rho} - \frac{\mu}{c\rho d^2\phi}\Phi + gz = \text{constant}$$

So

$$\Phi = \frac{K}{\mu}(p + z\rho g) + \text{constant}$$

Differentiating the above equation with respect to z, we have

$$v_z = \frac{K}{\mu}\left(\frac{\partial p}{\partial z} + \rho g\right),$$

where $K = cd^2\phi$. The above equation has the same form with Darcy's law, so Darcy's law can be regarded as the empirical formula of N-S equation.

6.2.3 Application Scope of Darcy's Law

Darcy's law is the basic law which comes from experiment using homogeneous sands, so it must have corresponding application scope. It can be seen from Darcy's law that the head loss is directly proportional to velocity, which is similar to laminar flow. Therefore, Darcy's law is only suitable for laminar flow or linear flow, the flow which exceeds this application scope is called non-Darcy flow.

When velocity increases to a certain value, there will be inertial resistance besides viscous resistance, then the flow rate and pressure difference will not have linear relationship, this velocity value is called the critical velocity of Darcy's law (Curve 1 in Fig. 6.3). If velocity exceeds the critical velocity, then the flow could not keep linear all the time changing to nonlinear, thus Darcy's law is no longer suitable. In Fig. 6.3, if the pressure gradient exceeds b, the flow is non-Darcy flow.

When fluid flows through low-permeability tight rock at low velocity, due to the adsorption effect between the fluid and rock, the fluid will only start flow if the pressure gradient exceeds the starting pressure gradient (point a in Fig. 6.3).

6.2.4 Determination of Permeability or Hydraulic Conductivity

Hydraulic conductivity k is a parameter which comprehensively reflects the flow ability of porous media, the correct determination of its value is of great

Fig. 6.3 Relationship between pressure gradient and velocity

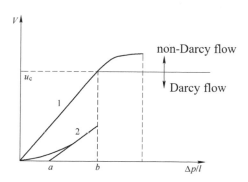

significance to calculation. There are many factors influencing hydraulic conductivity, mainly depending on fluid viscosity and the properties of solid particles, such as shape, size, and nonuniformity coefficient. Therefore, it is difficult to develop an exact theoretical formula to calculate hydraulic conductivity k, the common method is experiment or experience.

(1) Laboratory measure method: measure the heard loss and flow rate in the laboratory, and then use the theoretical formula to calculate the coefficient.
(2) Field method: drill a well and measure the flow rate and head loss, then use the theoretical formula to calculate the value of permeability or hydraulic conductivity. It can be obtained from Eq. (6.15) that: $K = \frac{Q\mu L}{A \Delta p}$.
(3) Experience method: In various relevant handbooks, there are values or formulas to calculate hydraulic conductivities of various kinds of soils. However, most of them are empirical and have limitations, so they can only be used for the rough estimate. Hydraulic conductivities of various kinds of soils are listed in Table 6.1 for reference.

Example 6.1 There are two vessels connected by a horizontal square tube as shown in Fig. 6.4. The dimensions are $a = 20$ cm and $l = 100$ cm. The tube is filled with coarse sand and its hydraulic conductivity is $k = 0.05$ cm/s. The water depths of two vessels are $H_1 = 80$ cm and $H_2 = 40$ cm, try to determine the flow rate of the tube. If the back half of the tube is replaced with fine sand (hydraulic conductivity $k = 0.005$ cm/s), try to determine the flow rate of the tube.
Solution

(1) When the tube is filled with coarse sand, according to Eq. (6.11), we have

$$Q = kAJ,$$

where $A = a^2$ and $J = \frac{H_1 - H_2}{l}$, so

$$Q = ka^2 \frac{H_1 - H_2}{l}$$

$$= 0.05 \times 20^2 \times \frac{80 - 40}{100} = 8\,(\text{cm}^3/\text{s}) = 0.008\,(\text{L/s})$$

(2) If the first half is coarse sand and the second half is fine sand, letting H denote the head of the cross section of the tube, then from Eq. (6.11), the flow rates of the coarse sand section and the fine sand section are respectively

$$Q_1 = k_1 \frac{H_1 - H}{0.5l} A \quad \text{and} \quad Q_2 = k_2 \frac{H - H_2}{0.5l} A$$

Table 6.1   Reference values of soil hydraulic conductivities

Soil type	Hydraulic conductivity k	
	m/day	cm/s
Clay	<0.005	$<6 \times 10^{-6}$
Loam	0.005–0.1	6×10^{-6}–1×10^{-4}
Light loam	0.1–0.5	1×10^{-4}–6×10^{-4}
Loess	0.25–0.5	3×10^{-4}–6×10^{-4}
Sandy silt	0.5–1.0	6×10^{-4}–1×10^{-3}
Fine sand	1.0–5.0	1×10^{-3}–6×10^{-3}
Medium sand	5.0–20.0	6×10^{-3}–2×10^{-2}
Homogenous medium sand	35–50	4×10^{-2}–6×10^{-2}
Coarse sand	20–50	2×10^{-2}–6×10^{-2}
Homogeneous coarse sand	60–75	7×10^{-2}–8×10^{-2}
Round gravel	50–100	6×10^{-2}–1×10^{-1}
Pebble	100–500	1×10^{-1}–6×10^{-1}
Pebble without filling	500–1000	6×10^{-1}–1×10
Slightly cracked pebbles	20–60	2×10^{-2}–7×10^{-2}
Multi fractured pebble	>60	$>7 \times 10^{-2}$

According to continuity principle we have $Q_1 = Q_2$, namely

$$k_1 \frac{H_1 - H}{0.5l} A = k_2 \frac{H - H_2}{0.5l} A$$

$$H = \frac{k_1 H_1 + k_2 H_2}{k_1 + k_2} = \frac{0.05 \times 80 + 0.005 \times 40}{0.05 + 0.005} = 76.36 \,(\text{cm})$$

The flow rate is

$$Q = Q_1 = k_1 \frac{H_1 - H}{0.5l} A = 0.05 \times \frac{80 - 76.36}{0.5 \times 100} \times 20^2 = 1.456 \,(\text{cm}^3/\text{s})$$

6.3   Mathematical Model of Liquid Porous Flow

The research steps are usually divided into four steps

(1) Make a reasonable abstraction and simplification of complex practical problem, and establish an ideal physical model;
(2) Establish corresponding mathematical model for the physical model;

Fig. 6.4 Connected vessels

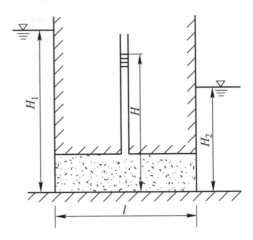

(3) Solve the mathematical model;
(4) Apply the theoretical solution to actual problem, find the gap between the theoretical solution and actual problems and correct the established models to make them closer to actual problems.

Repeat above steps until obtaining desirable results, as shown in Fig. 6.5.

6.3.1   Mathematical Model of Fluid Flow Through Porous Media

A whole mathematical model consists of two parts: one is basic differential equation, the other is boundary or initial condition. To model the basic rule of fluid flow through porous media, the following factors must be considered:

(1) Mass conservation law is a universal law in nature, so the basic differential equation must be based on the continuity equation which describes mass conservation law.
(2) Fluid flow through porous media is a kind of fluid motion, so it must be dominated by motion equation.
(3) Fluid flow through porous media is also influenced by the fluid and rock states, so it is necessary to establish state equations.
(4) Fluid flow through porous media is sometimes accompanied by some physical and chemical phenomena, so characteristic equation should be established to describe these special phenomena.

Fig. 6.5 General research
steps [6]

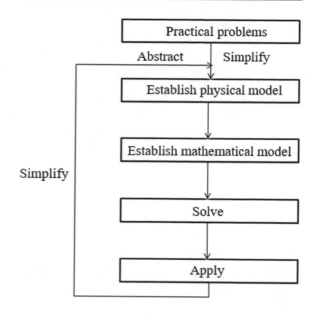

6.3.1.1   Continuity Equation

Take a cubic element $AA'B'B$ from the formation as shown in Fig. 6.6, its
dimensions are dx, dy and dz. The mass velocity of the center M is $\rho(p)v$, its
components are $\rho(p)v_x$, $\rho(p)v_y$ and $\rho(p)v_z$, where $\rho(p)$ is the liquid density.

In the x direction, the component of mass velocity of particle M_A is:

$$\rho(p)v_x - \frac{\partial[\rho(p)v_x]}{\partial x} \cdot \frac{dx}{2} \tag{6.16}$$

After time dt, the mass passing through surface AA' is

$$\left\{ \rho(p)v_x - \frac{\partial[\rho(p)v_x]}{\partial x} \cdot \frac{dx}{2} \right\} dydzdt \tag{6.17}$$

Similarly, the component of mass velocity of particle M_B is

$$\rho(p)v_x + \frac{\partial[\rho(p)v_x]}{\partial x} \cdot \frac{dx}{2} \tag{6.18}$$

After time dt, the mass passing through surface BB' is

$$\left\{ \rho(p)v_x + \frac{\partial[\rho(p)v_x]}{\partial x} \cdot \frac{dx}{2} \right\} dydzdt \tag{6.19}$$

Fig. 6.6 Cubic element

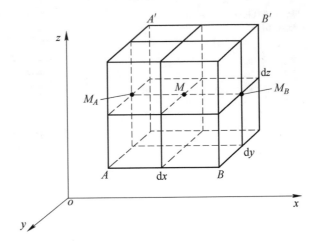

The mass difference between the inflow and outflow of the cubic element during time dt in x direction is

$$-\frac{\partial[\rho(p)v_x]}{\partial x}dxdydzdt \qquad (6.20)$$

Similarly, the mass differences between the inflow and outflow of the cubic element during time dt in y and z direction are

In the y direction $-\frac{\partial[\rho(p)v_y]}{\partial y}dxdydzdt$

In the z direction $-\frac{\partial[\rho(p)v_z]}{\partial z}dxdydzdt$

So the total mass difference between the inflow and outflow of the cubic element can be obtained

$$-\left\{\frac{\partial[\rho(p)v_x]}{\partial x}+\frac{\partial[\rho(p)v_y]}{\partial y}+\frac{\partial[\rho(p)v_z]}{\partial z}\right\}dxdydzdt \qquad (6.21)$$

Equation (6.21) should equal the fluid mass change in the cubic element during time dt. The liquid mass change during time dt is

$$\frac{\partial[\rho(p)\phi]}{\partial t}dxdydzdt \qquad (6.22)$$

Thus the following equation can be obtained:

$$-\left[\frac{\partial[\rho(p)v_x]}{\partial x}+\frac{\partial[\rho(p)v_y]}{\partial y}+\frac{\partial[\rho(p)v_z]}{\partial z}\right]=\frac{\partial[\rho(p)\phi]}{\partial t} \qquad (6.23)$$

Or

$$-\nabla \cdot [\rho(p)\mathbf{v}] = \frac{\partial[\rho(p)\phi]}{\partial t}$$

Equation (6.23) is the continuity equation of unsteady flow through porous media for slightly compressible liquid.

6.3.1.2   Motion Equation

For linear flow, its motion equation obeys Darcy's law. In rectangular coordinate system Darcy's law can be written as

$$v_x = -\frac{K\partial p}{\mu \partial x}$$

$$v_y = -\frac{K\partial p}{\mu \partial y}$$

$$v_z = -\frac{K\partial p}{\mu \partial z}$$

It can be written in vector form

$$\mathbf{v} = -\frac{K}{\mu}\nabla p \qquad (6.24)$$

6.3.1.3   State Equations

(1)   State equation of liquid

The main parameter used to describe liquid elastic state is liquid compression coefficient C_L

$$C_L = \frac{-\frac{dv}{V}}{dp} \qquad (6.25)$$

Liquid density ρ is relevant to mass m and volume V

$$V = \frac{m}{\rho} \qquad (6.26)$$

Differentiating both sides of Eq. (6.26), we have

$$dV = md\left(\frac{1}{\rho}\right) = -m\rho^{-2}d\rho \qquad (6.27)$$

Substituting V and dV into Eq. (6.25), we obtain

$$C_L = \frac{\frac{d\rho}{\rho}}{dp} \qquad (6.28)$$

Separating variables and integrating, we have

$$C_L \int_{p_a}^{p} dp = \int_{\rho_a}^{\rho} \frac{d\rho}{\rho}$$

$$C_L(p - p_a) = \ln\frac{\rho}{\rho_a}$$

$$\rho = \rho_a e^{C_L(p-p_a)} \qquad (6.29)$$

The exponential function can be expanded in series

$$e^{C_L(p-p_a)} = 1 + C_L(p - p_a) + \frac{C_L^2}{2!}(p - p_a)^2 + \cdots$$

Since C_L is very small, the higher order terms can be neglected, thus

$$\rho = \rho_a[1 + C_L(p - p_a)] \qquad (6.30)$$

Equation (6.30) is the expression for slightly compressible fluid density for unsteady flow. For steady flow, fluid is incompressible and its density ρ is constant.

(2) State equation of rock

Rock elastic compression coefficient C_f is used to describe its elastic state

$$C_f = \frac{\frac{dv_p}{V_p}}{dp}, \qquad (6.31)$$

where V_p is the void volume.

$$\frac{dV_p}{V_p} = \frac{d\phi}{\phi}$$

$$C_f = \frac{1}{\phi}\frac{d\phi}{dp} \qquad (6.32)$$

Integrating Eq. (6.32), we have

$$C_f \int_{p_a}^{p} dp = \int_{\phi_a}^{\phi} \frac{1}{\phi} d\phi$$

$$C_f(p - p_a) = \ln\left(\frac{\phi}{\phi_a}\right)$$

$$\phi = \phi_a e^{C_f(p-p_a)} \tag{6.33}$$

Equation (6.33) can be expanded in series and the higher order terms can be neglected

$$\phi = \phi_a[1 + C_f(p - p_a)] \tag{6.34}$$

Equation (6.34) is the expression for slightly compressible rock porosity for unsteady flow. For steady flow, porous medium is incompressible and its porosity ϕ is constant.

6.3.2   Basic Differential Equation

Substituting motion equation and state equations into the continuity equation, then the basic differential equation can be obtained. Now substituting Eqs. (6.24), (6.29) and (6.33) into Eq. (6.23), then the left hand of Eq. (6.23) consists of three terms, take one term for example

$$
\begin{aligned}
\frac{\partial(\rho v_x)}{\partial x} &= \frac{\partial}{\partial x}\left[\rho_a e^{C_L(p-p_a)}\left(-\frac{K\partial p}{\mu \partial x}\right)\right] \\
&= -\frac{K}{\mu}\rho_a \frac{\partial}{\partial x}\left[e^{C_L(p-p_a)}\frac{\partial p}{\partial x}\right] \\
&= -\frac{K}{\mu}\rho_a \frac{\partial}{\partial x}\left[\frac{\partial}{\partial x}\left(\frac{e^{C_L(p-p_a)}}{C_L}\right)\right] \\
&= -\frac{K}{\mu}\frac{\rho_a}{C_L}\frac{\partial}{\partial x}\left[\frac{\partial}{\partial x}e^{C_L(p-p_a)}\right] \\
&= -\frac{K}{\mu}\frac{\rho_a}{C_L}\frac{\partial}{\partial x}\left\{\frac{\partial}{\partial x_a}[1 + C_L(p - p_a)]\right\} \\
&= -\frac{K}{\mu}\frac{\rho_a}{C_L}C_L\frac{\partial^2 p}{\partial x^2} \\
&= -\frac{K}{\mu}\rho_a \frac{\partial^2 p}{\partial x^2}
\end{aligned}
$$

Similarly

$$\frac{\partial(\rho v_y)}{\partial y} = -\frac{K}{\mu}\rho_a\frac{\partial^2 p}{\partial y^2}$$

$$\frac{\partial(\rho v_z)}{\partial z} = -\frac{K}{\mu}\rho_a\frac{\partial^2 p}{\partial z^2}$$

So

$$\frac{\partial(\rho v_x)}{\partial x} + \frac{\partial(\rho v_y)}{\partial y} + \frac{\partial(\rho v_z)}{\partial z} = -\frac{K}{\mu}\rho_a\left(\frac{\partial^2 p}{\partial y^2} + \frac{\partial^2 p}{\partial y^2} + \frac{\partial^2 p}{\partial z^2}\right) \tag{6.35}$$

Simplifying the right hand of the Eq. (6.23), then $\rho \cdot \phi$ can be expressed as

$$\rho \cdot \phi = \rho_a[1 + C_L(p - p_a)] \cdot \phi_a[1 + C_f(p - p_a)]$$
$$= \rho_a\phi_a + \rho_a\phi_a(C_L + C_f)(p - p_a) + \rho_a\phi_a C_L C_f(p - p_a)^2$$

Since $C_L C_f$ is very small, so $C_L C_f$ is neglected. Define total compression coefficient as

$$C_t = C_L + C_f$$

Then $\rho \cdot \phi = \rho_a\phi_a + \rho_a\phi_a C_t(p - p_a)$

$$\frac{\partial(\rho\phi)}{\partial t} = \rho_a\phi_a C_t\frac{\partial p}{\partial t} \tag{6.36}$$

Substituting Eqs. (6.35) and (6.36) into Eq. (6.23) and replacing ϕ_a with ϕ for convenience, we have

$$\frac{K}{\phi\mu C_t}\left(\frac{\partial^2 p}{\partial x^2} + \frac{\partial^2 p}{\partial y^2} + \frac{\partial^2 p}{\partial z^2}\right) = \frac{\partial p}{\partial t} \tag{6.37}$$

Setting $\frac{K}{\phi\mu C_t} = \eta$, which is called pressure transmitting coefficient. Its physical meaning is the area of pressure transmission per unit time and its dimension is m^2/s. So Eq. (6.37) can be written as

$$\eta\left(\frac{\partial^2 p}{\partial x^2} + \frac{\partial^2 p}{\partial y^2} + \frac{\partial^2 p}{\partial z^2}\right) = \frac{\partial p}{\partial t} \tag{6.38}$$

or

$$\eta \nabla^2 p = \frac{\partial p}{\partial t}$$

This is the basic differential equation of unsteady flow for slightly compressible liquid in porous media, which is the theoretical basis for solving unsteady flow of slightly compressible liquid.

The physical model of single liquid steady flow in porous media studied in this chapter is: the formation is homogeneous horizontal incompressible and isotropic; the liquid is single phase incompressible Newtonian fluid. The flow process is also assumed to be isothermal steady without any physical and chemical phenomena and obey Darcy's law.

The mathematical model has been established, now specific conditions to solve this model are also needed. In this section two cases are solved: one is planar one-directional flow, the other is planar radial flow, and the pressure distributions and flow rate formulas of both cases are also obtained.

For steady flow, the inflow mass equals outflow mass and the fluid density is constant. So the right-hand side of Eq. (6.38) equals 0. According to assumptions, variables K, μ and ρ are constant, so

$$\frac{\partial^2 p}{\partial x^2} + \frac{\partial^2 p}{\partial y^2} + \frac{\partial^2 p}{\partial z^2} = 0 \qquad (6.39)$$

The above equation is the basic differential equation of steady flow for single phase liquid, which is also called Laplace equation and can be written as

$$\nabla^2 p = 0$$

6.4 Solution for Planar One-Directional Flow

The simplified physical model is shown in Fig. 6.7, assuming the formation is homogeneous and horizontal, and its permeability is K. One end of the formation is supply boundary, and its pressure is p_e; the other end is the drainage channel, and its pressure is p_B. The formation length is L, its width is W and its depth is h. The Newtonian fluid flows along x direction and its viscosity is μ.

Fig. 6.7 Simplified model of
planar one-directional flow

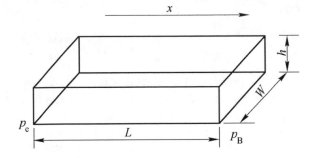

(1)　Pressure distribution

Assuming the liquid only flows along x direction, then the basic differential
equation can be simplified to one-dimensional form

$$\frac{\mathrm{d}^2 p}{\mathrm{d}x^2} = 0 \tag{6.40}$$

Boundary conditions are

$$x = 0 \quad p = p_e$$

$$x = L \quad p = p_B \tag{6.41}$$

Integrating Eq. (6.40), we have

$$\frac{\mathrm{d}p}{\mathrm{d}x} = C_1 \tag{6.42}$$

Integrating Eq. (6.42), we have

$$p = C_1 x + C_2, \tag{6.43}$$

where C_1 and C_2 are integral constants.
　　Substituting boundary conditions (6.41) into Eq. (6.43), we obtain

$$\begin{cases} C_2 = p_e \\ C_1 = -\frac{p_e - p_B}{L} \end{cases} \tag{6.44}$$

　　Substituting C_1 and C_2 into Eq. (6.43), the pressure distribution of every point in
the formation can be obtained

$$p = p_e - \frac{p_e - p_B}{L} x \tag{6.45}$$

Fig. 6.8 Pressure
distribution of planar
unidirectional flow

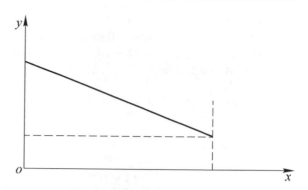

The pressure distribution curve of one-directional steady flow can be obtained from Eq. (6.45), as shown in Fig. 6.8. The curve shows that the pressure is in linear distribution from supply boundary to drainage channel, its slope is $-\frac{p_e - p_B}{L}$, which means the pressure decreases evenly.

(2) Flow rate formula

The flow rate formula can be obtained according to Darcy's law $v = -\frac{K dp}{\mu dx}$:

$$q = A \cdot v = -\frac{K}{\mu} Wh \frac{dp}{dx}, \tag{6.46}$$

where A is flow area, $A = Wh$; W is the formation width; h is the formation depth. From Eq. (6.45), we have

$$\frac{dp}{dx} = -\frac{p_e - p_B}{L} \tag{6.47}$$

Substituting Eq. (6.47) into Eq. (6.46), the flow rate formula of one-directional flow can be obtained

$$q = \frac{KWh(p_e - p_B)}{\mu L} \tag{6.48}$$

Example 6.2 An experiment is carried out to measure the permeability of a cylindrical core. The core radius is 1 cm, and the core length is 5 cm. The liquid flows through the core driven by the pressure difference between the two ends, and the viscosity of the liquid is 1×10^3 Pa · s. The volume of the liquid passing through the core during 2 min is 15 cm³. The pressure difference of the two ends is 157 mm Hg known from the mercury manometer. Try to determine the permeability K of the core.

Solution

It is known that $r = 0.01$ m, $L = 0.05$ m, $\mu = 1 \times 10^3$ Pa $\cdot$ s, $t = 2 \times 60 = 120$ s, $h = 157 \times 10^{-3}$ m and $V = 15 \times 10^{-6}$ m^3, the pressure difference can be calculated by density ρ and h:

$$\Delta p = \rho g h = 13.6 \times 10^3 \times 9.8 \times 157 \times 10^{-3} = 20924.96 \text{ Pa}$$

The flow rate is

$$q = \frac{V}{t} = \frac{15 \times 10^{-6}}{120} = 0.125 \times 10^{-6} \text{ m}^3/\text{s}$$

It can be obtained from Darcy's law that

$$K = \frac{q\mu L}{A\Delta p} = \frac{0.125 \times 10^{-6} \times 10^3 \times 0.05}{3.14 \times 0.01^2 \times 20924.96} = 9.5 \times 10^{-7} \text{ m}^2$$

So the permeability of the core is 9.5×10^{-7} m^2.

(3) Flow field chart of planar one-directional flow

The chart which consists of a group of isobars and streamlines according to a certain rule is called flow field chart. The isobar is referred to the line on which every point has the same pressure, and the line which is vertical to the isobar is called streamline.

It can be known from Eq. (6.45) that the pressure of all points with same x coordinate is identical. So the isobars are parallel to y axis and the streamlines are parallel to x axis, they have formed a uniform network, as shown in Fig. 6.9.

6.5 Solution for Planar Radial Flow

The simplified physical model is shown in Fig. 6.10, assuming the formation is a homogeneous and horizontal disk, its permeability is K and its thickness is h. The outer circle boundary is the supply boundary, its pressure is p_e and the supply radius is r_e. A well is drilled at the circle center, its radius is r_w and its pressure is p_{wf}. The liquid is Newtonian fluid and its viscosity is μ.

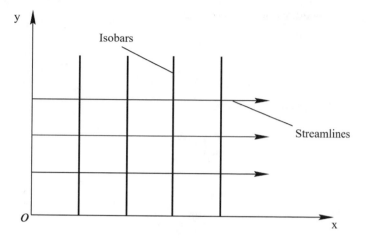

Fig. 6.9 Flow field chart of planar one-directional flow

Fig. 6.10 Simplified model of planar radial flow

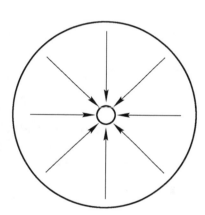

(1) Pressure distribution

Planar radial flow is a two-dimensional flow, so Eq. (6.39) can be simplified as

$$\frac{\partial^2 p}{\partial x^2} + \frac{\partial^2 p}{\partial y^2} = 0 \tag{6.49}$$

Its polar form is

$$\frac{d^2 p}{dr^2} + \frac{1}{r}\frac{dp}{dr} = 0 \tag{6.50}$$

Boundary conditions of the model are

$$r = r_e, \quad p = p_e$$

$$r = r_w, \quad p = p_{wf} \tag{6.51}$$

Equation (6.39) can be written as

$$\frac{d}{dr}\left(r\frac{dp}{dr}\right) = 0 \tag{6.52}$$

By integrating

$$r\frac{dp}{dr} = C_1 \tag{6.53}$$

By separating variables

$$dp = C_1 \frac{1}{r}dr \tag{6.54}$$

By integrating

$$p = C_1 \ln r + C_2 \tag{6.55}$$

Substituting boundary conditions (6.51) into Eq. (6.55), C_1 and C_2 can be obtained, then substituting C_1 and C_2 into Eq. (6.55), the pressure distribution of every point in the formation can be obtained

$$p = p_e - \frac{p_e - p_{wf}}{\ln \frac{r_e}{r_w}} \ln \frac{r_e}{r} \tag{6.56}$$

It can be known from Eq. (6.56) that the pressure has a logarithmic distribution from the supply boundary to well bottom, as shown in Fig. 6.11. Its shape is similar to a funnel, so it is also called "pressure drop funnel." The pressure is mainly consumed near the well bottom, which is because the smaller flow area near well bottom, the greater resistance to flow.

(2) Flow field chart

It can be known from Eq. (6.56) that the pressure of all points with same r coordinate is identical, so the isobars are a group of concentric circles and the streamlines are a group of radial rays. The closer to well bottom, the denser the lines as shown in Fig. 6.12.

Fig. 6.11 Pressure distribution of planar radial flow [6]

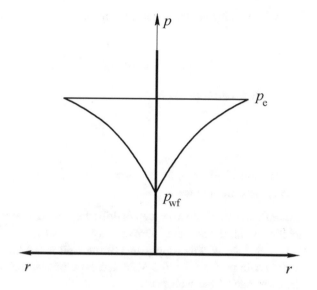

Fig. 6.12 Streamline chart of planar radial flow [6]

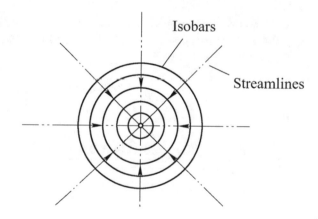

(3)   Flow rate formula

It can be known from Darcy's law that

$$q = 2\pi rh \frac{K}{\mu} \left(\frac{\mathrm{d}p}{\mathrm{d}r} \right) \tag{6.57}$$

So

$$\frac{q\mu}{2\pi hK} \frac{1}{r} \mathrm{d}r = \mathrm{d}p \tag{6.58}$$

By integrating on both sides, we have

$$\frac{q\mu}{2\pi hK} \int_{r_w}^{r_e} \frac{1}{r} dr = \int_{p_w}^{p_e} dp$$

The flow rate formula is

$$q = \frac{2\pi Kh(p_e - p_w)}{\mu \ln \frac{r_e}{r_w}} \qquad (6.59)$$

The above equation is the flow rate formula of planar radial flow, which is widely used in engineering.

Example 6.3 A productive well is drilled by using gravity water pressure, the water thickness is 10 m, the permeability is 0.4×10^{-12} m^2, water viscosity is 1.5×10^{-3} Pa · s. The supply area of the well is 0.3 km^2, well radius is 0.1 m. The water static pressure is 10.5 MPa and flow pressure is 7.5 MPa. Try to determine the flow rate of the well per day.

Solution

It is known that $A = 0.3$ km^2, $r_w = 0.1$ m, $p_e = 10.5$ MPa, $p_w = 7.5$ MPa, $K = 0.4 \times 10^{-12}$ m^2, $\mu = 1.5 \times 10^{-3}$ Pa · s and $h = 10$ m.

Since $A = \pi r_e^2 = 0.3$ km^2,

$$r_e = 309 \text{ m}$$

From Eq. (6.59), we have

$$q = \frac{2\pi Kh(p_e - p_{wf})}{\mu \ln \frac{r_e}{r_w}} = \frac{2 \times 3.14 \times 0.4 \times 10^{-12} \times 10 \times (10.5 - 7.5) \times 10^6}{1.5 \times 10^{-3} \times \ln \frac{309}{0.1}}$$

$$= 0.625 \times 10^{-2} \text{ m}^3/\text{s}$$

6.6 Problems

6.1. To use Darcy's law to measure hydraulic conductivity of some kind of soil by experiment, put the soil sample in a tube which's diameter is $D = 30$ cm. The amount of flow for 6 h is 85L and the action head is 80 cm. The distance between the two piezometers is 40 cm, so the hydraulic conductivity of the soil is

(A) 0.4 m/d   (B) 1.4 m/d   (C) 2.4 m/d   (D) 3.4 m/d

6.2. The flow rate of the tubular formation model is $Q = 12$ cm^3/min, the model diameter is $D = 2$ cm. The viscosity of the used liquid is $\mu = 9 \times 10^{-3}$ Pa · s and its density is 0.85 g/cm^3. The model porosity is $\phi = 0.2$ and the permeability is $K = 1 \times 10^{-12}$ m^2. Try to determine the flow velocity v and real velocity u of the liquid.

6.3. A pumping test is carried out in a homogeneous phreatic aquifer to determine hydraulic conductivity k. The aquifer thickness is 12 m, the well diameter is 20 cm. The distance between the well and the observation well is 20 m. When the pump keeps constant as 2L/s, the water level of the well decreases by 2.5 m and the water level of the observation well decreases by 0.38 m. Try to determine the value of k.

6.4. A core is positioned at an angle as shown in Fig. 6.13 and its length is 100 cm. The section of the core is a square and the square side length is 5 cm. The inlet pressure is $p_1 = 0.2$ MPa and the outlet pressure is $p_2 = 0.1$ MPa. $h = 50$ cm, the specific weight of the liquid is 0.85 times of that of water. The length of the flow section is $L = 100$ cm, the liquid viscosity is $\mu = 2 \times 10^{-3}$ Pa · s, the rock permeability is $K = 1 \times 10^{-12}$m^2. Try to determine the flow rate Q.

6.5. Fig. 6.14 is a completed well, and the impermeable layer is flat. The well radius is $r_o = 10$ m, the thickness of the aquifer is $H = 8$ m. The soil is fine sand and its hydraulic conductivity is $k = 0.001$ m/s. Try to calculate the maximum water flow rate when the water depth of the well $h_0 \geq 2$ m, and determine the relationship between the water level of the well and water flow rate.

6.6. Fig. 6.15 is an overflow dam. The water level elevation at the upstream of the dam is 125 m and that of the downstream is 105 m. Water seeps through the dam, and the flow net is shown in Fig. 6.15. The hydraulic conductivity is $k = 5 \times 10^{-5}$ m/s. The water level elevations of point A and B are 92 and 97 m, respectively. At point C, $\Delta s = \Delta b = 3$ m.

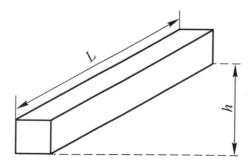

Fig. 6.13 Problem 6.4

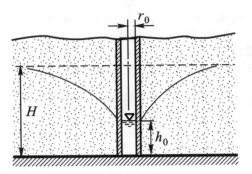

Fig. 6.14 Problem 6.5

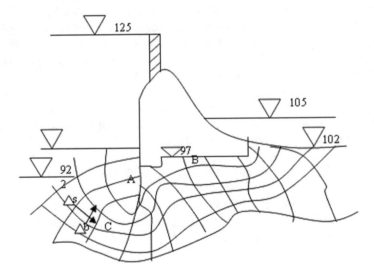

Fig. 6.15 Problem 6.6

Try to calculate

(1) The flow velocity of point C under the dam;
(2) The flow rate per unit width of the dam base;
(3) The flow pressure of Point A and B at the dam bottom.

References

1. Athy, L.F.: Density, porosity, and compaction of sedimentary rocks. AAPG Bulletin **14**(1), 1–24 (1930)
2. Xie, Z.: Engineering fluid mechanics, 4th edn. Metallurgical Industry Press, Beijing (2014)
3. Darcy, H.: Les fontaines publiques de la ville de Dijon, Victor Dalmont (1856)
4. Whitaker, S.: Flow in porous media i: a theoretical derivation of Darcy's law. Transp. Porous Media **1**(1), 3–25 (1986)
5. Klinkenberg, L.J.: The permeability of porous media to liquids and gases. In Drilling and production practice, American Petroleum Institute (1941)
6. Song, H.: Engineering fluid mechanics and environmental application. Metallurgical Industry Press, Beijing (2016)

Fluid Machinery

7

Abstract

Pumps and fans all belong to fluid machinery which can be used to transport fluids. Pumps are always used to transport liquids and fans are always used to transport gases. In this chapter, we will first discuss the characteristics of centrifugal pump, including its head rise, efficiency, performance curves. Then we will introduce system curve and pump selection. Finally, we will indicate some basic concepts about centrifugal fan.

Keywords

Centrifugal pump · Head rise · Efficiency · Operating point · Centrifugal fan

7.1 Centrifugal Pump

Centrifugal pumps are a sub-class of dynamic axisymmetric work-absorbing fluid machinery [1, 2]. Centrifugal pumps are used to transport fluids by the conversion of rotational kinetic energy to the hydrodynamic energy of the fluid flow. The rotational energy typically comes from an engine or electric motor. As shown in Fig. 7.1, fluid enters axially through eye of the casing, is caught up in the impeller blades, and is whirled tangentially and radially outward until it leaves through all circumferential parts of the impeller into the diffuser part of the casing. The fluid gains both velocity and pressure while passing through the impeller. The doughnut-shaped diffuser, or scroll, section of the casing decelerates the flow and further increases the pressure.

© Metallurgical Industry Press, Beijing and Springer Nature Singapore Pte Ltd. 2018   199
H. Song, *Engineering Fluid Mechanics*,
https://doi.org/10.1007/978-981-13-0173-5_7

Fig. 7.1 Centrifugal pump
[3], (1-impeller 2-impeller
blades 3-suction side 4-casing
5-outlet)

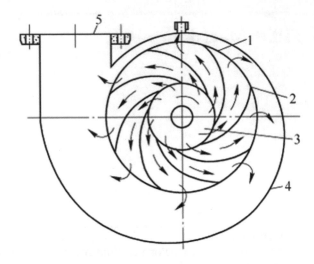

7.1.1 Theoretical Considerations

7.1.1.1 Head Rise of the Pump

As shown in Fig. 7.2, section 1-1 is the inlet of the pump with vacuum gauge 3, and
section 2-2 is the outlet of the pump with pressure gauge 4. The energy difference

Fig. 7.2 Centrifugal pump in
a piping system [3]

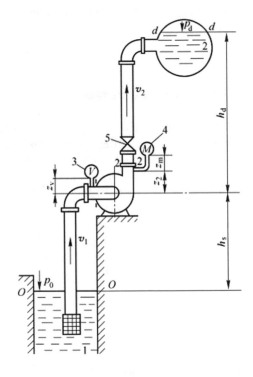

of per unit weight fluid between outlet and inlet $e_2 - e_1$, or the energy head added to the flow, is called head rise of the pump, which is denoted by H.

Namely $\qquad H = e_2 - e_1$

As shown in Fig. 7.2, based on the surface O–O of the reservoir 1, the energy of per unit weight fluid in sections 1-1 and 2-2 is

$$e_1 = h_s + \frac{p_1}{\gamma} + \frac{v_1^2}{2g}$$

$$e_2 = h_s + z_2 + \frac{p_2}{\gamma} + \frac{v_2^2}{2g},$$

where γ is the specific weight of the liquid.

Assuming that the atmospheric pressure is p_a, and vacuum gauge indicates p_v, pressure gauge indicates p_M, then

$$p_1 = p_a - p_v + \gamma z_v$$

$$p_2 = p_a + p_M + \gamma z_m$$

Thus

$$H = e_2 - e_1$$

$$= h_s + z_2 + \frac{p_a + p_M}{\gamma} + z_m - h_s - \frac{p_a - p_v}{\gamma} - z_v + \frac{v_2^2 - v_1^2}{2g}$$

Namely

$$H = (z_2 + z_m) - z_v + \frac{p_M + p_v}{\gamma} + \frac{v_2^2 - v_1^2}{2g} \qquad (7.1)$$

$(z_2 + z_m) - z_v = \Delta z$ is the elevation height difference between pressure gauge and vacuum gauge, which is always small and can be neglected. Usually v_1 and v_2 are about the same, then the head rise

$$H = \frac{p_M + p_v}{\gamma} \qquad (7.2)$$

Therefore, the head rise of the pump can be calculated by pressure gauge and vacuum gauge.

According to Fig. 7.2, based on section O–O, we write Bernoulli equation between the surface of the reservoir 1 and section 1-1 as

$$\frac{p_0}{\gamma} + \frac{v_0^2}{2g} = h_s + \frac{p_1}{\gamma} + \frac{v_1^2}{2g} + h_{ls} \tag{7.3}$$

Then $\quad e_1 = h_s + \frac{p_1}{\gamma} + \frac{v_1^2}{2g} = \frac{p_0}{\gamma} + \frac{v_0^2}{2g} - h_{ls}$

we write Bernoulli equation between section 2-2 and the surface d-d of reservoir 2 as:

$$h_s + z_2 + \frac{p_2}{\gamma} + \frac{v_2^2}{2g} = h_s + h_d + \frac{p_d}{\gamma} + \frac{v_d^2}{2g} + h_{ld} \tag{7.4}$$

Then $\quad e_2 = h_s + h_d + \frac{p_d}{\gamma} + \frac{v_d^2}{2g} + h_{ld}$

Thus

$$H = e_2 - e_1$$

$$= h_s + h_d + \frac{p_d}{\gamma} + \frac{v_d^2}{2g} + h_{ld} - \frac{p_0}{\gamma} - \frac{v_0^2}{2g} + h_{ls}$$

Because of the large area of reservoir 1 and reservoir 2 surface, $v_d \approx 0$, $v_0 \approx 0$, thus

$$H = h_s + h_d + h_{ls} + h_{ld} + \frac{p_d - p_0}{\gamma} \tag{7.5}$$

When the reservoir 1 and reservoir 2 are connected with the atmosphere, $p_d = p_a = p_0$, then the head rise of the pump

$$H = h_s + h_d + h_{ls} + h_{ld} = H_G + H_l, \tag{7.6}$$

where H_G represents the lift height of the liquid, and H_l is the head loss in the pipe.

7.1.1.2 Efficiency of the Pump

The power delivered to the fluid simply equals the specific weight times the flow rate times the head rise of the pump

$$P_w = \gamma Q H \tag{7.7}$$

This is traditionally called output power. The power required to drive the pump is called input power, denoted by N

$$N = \omega T, \tag{7.8}$$

where ω is the shaft angular velocity and T is the shaft torque. If there are no losses, P_w and N will be equal, but of course P_w is actually less, and the efficiency η of the pump is defined as

$$\eta = \frac{P_w}{N} = \frac{\gamma Q H}{N} \qquad (7.9)$$

The efficiency η is basically composed of three parts: volumetric, hydraulic, and mechanical. The volumetric efficiency is

$$\eta_v = \frac{Q}{Q + Q_L}, \qquad (7.10)$$

where Q_L is the loss of fluid due to leakage in the impeller casing clearances. The hydraulic efficiency is

$$\eta_h = 1 - \frac{h_f}{h_s}, \qquad (7.11)$$

where h_f has three parts: (1) shock loss at the eye due to imperfect match between inlet flow and the blade entrances, (2) friction losses in the blade passages, and (3) circulation loss due to imperfect match at the exit side of the blades. h_s is called theoretical head rise of the pump.

Finally, the mechanical efficiency is

$$\eta_m = 1 - \frac{P_f}{N}, \qquad (7.12)$$

where P_f is the power loss due to mechanical friction in the bearings, packing glands, and other contact points in the machine.

By definition, the total efficiency is simply the product of its three parts

$$\eta = \eta_v \, \eta_h \, \eta_m \qquad (7.13)$$

7.1.2 Pump Performance Characteristics

According to test data, knowing the flow rate Q for a centrifugal pump driven at constant speed, we can obtain the corresponding head rise H and input power N. The efficiency η of the pump can also be calculated according to Eq. (7.9). We can plot H, N, and η as functions of Q, and the curves are called performance curves for a centrifugal pump, as shown in Fig. 7.3. Note that there are different performance curves for different pumps with various rotation speed, but their trend and shape are almost similar to each other.

(1) *H-Q* curve. The head rise increases slowly to a peak as flow rate increases, then drops off at larger flow rate. The head loss is the main reason for the phenomena.

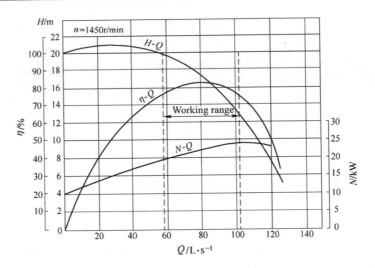

Fig. 7.3 Performance curves for a centrifugal pump

(2) N-Q curve. The input power rises with flow rate. In order to prevent large starting electric current from destroying the motor, the centrifugal pump always starts to work when $Q = 0$.

(3) η-Q curve. The efficiency is zero at shutoff (zero flow rate), and rises to a peak as flow rate increases, then drops off at larger flow rate. It stays near its maximum over a range of flow rate (in this example, from about 60–100 L/s). To make better use of the pump, the pump should work in such a "working range."

7.1.3 Net Positive Suction Head

Cavitation can occur in any machine-handling liquid whenever the local static pressure falls below the vapor pressure of the liquid. When this occurs, the liquid can locally flash to vapor, forming a vapor cavity and significantly changing the flow pattern from the noncavitating condition.

As cavitation commences, it reduces the performance of a pump rapidly. Thus, cavitation must be avoided to maintain stable and efficient operation. In addition, local surface pressures may become high when the vapor cavity implodes or collapses, causing erosion damage or surface pitting. In a pump, cavitation tends to begin at the section where the flow is accelerated into the impeller. Cavitation can be avoided if the pressure everywhere in the machine is kept above the vapor pressure of the operating liquid. At constant speed, this requires that a pressure somewhat greater than the vapor pressure of the liquid be maintained at a pump inlet (the suction).

Net positive suction head (NPSH) is defined as the difference between the absolute stagnation pressure in the flow at the pump suction and the liquid vapor pressure, expressed as head of flowing liquid. Hence the NPSH is a measure of the difference between the maximum possible pressure in the given flow and the pressure at which the liquid will start flashing over to a vapor; the larger the NPSH, the less likely cavitation is to occur. The net positive suction head required (NPSHR) by a specific pump to suppress cavitation varies with the liquid pumped, and with the liquid temperature and pump condition (e.g., as critical geometric features of the pump are affected by wear). The net positive suction head available (NPSHA) at the pump inlet must be greater than the NPSHR to suppress cavitation. Hence, for any inlet system, there is a flow rate that cannot be exceeded if flow through the pump is to remain free from cavitation. Inlet pressure losses may be reduced by increasing the diameter of the inlet piping; for this reason, many centrifugal pumps have larger flanges or couplings at the inlet than at the outlet.

7.1.4 System Characteristics and Pump Selection

7.1.4.1 Operating Point of the Pump in Fluid Systems

We define a fluid system as the combination of a fluid machine (like the pump) and a network of pipes or channels that convey fluid. As shown in Fig. 7.2, when the lift height of per unit weight liquid is H_G from the reservoir 1 to reservoir 2, the energy needed is H_A, thus

$$H_A = H_G + h_l$$

According to the equation in chapter 4,

$$
\begin{aligned}
h_l &= (\lambda \sum \frac{L}{D} + \sum \zeta) \frac{v^2}{2g} \\
&= (\lambda \sum \frac{L}{D} + \sum \zeta) \frac{Q^2}{2gA^2} = RQ^2,
\end{aligned}
\tag{7.14}
$$

where R is the constant determined only by the pipe system, thus

$$H_A = H_G + RQ^2 \tag{7.15}$$

Equation (7.15) is called system curve, and it is a parabolic curve, as shown in Fig. 7.4.

The intersection of the system curve with the pump performance curve H (Q) defines the operating point. Assuming that the peak of the H-Q curve is point F. On its right side, the operating point belongs to stable operating area. When the system curve becomes steeper because of the large H_G and R (for example, the long

Fig. 7.4 System curve

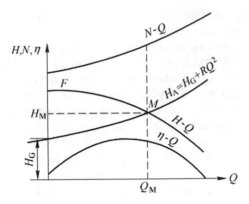

pipe with rough pipe wall and valves), the operating point may fall into the left side of point F. Correspondingly, the pump will work in an unstable way, sometimes even vibrate.

7.1.4.2 Pump Selection

First and most importantly, we have to take a close look at the application of the pump. There should be as much details about the system available as possible, to ensure the right pump could be chosen. Important selection parameters are required, such as head rise, flow rate, NPSHA, and so on. It is also important to know the properties of fluid, such as viscosity, abrasives, and temperature. To select a pump, it is useful to plot the curve, characterizing the system and the characteristic curve of a potential pump into the same diagram. The point of intersection of the two curves indicates the operating point of the pump in the system. It is also possible to make predictions how the pump will behave when changing system parameters. A common rule when selecting a pump is to choose a pump with at least 25% more head available than required by the system. Another common practice is to choose at most the second largest impeller diameter available in a pump series. In case of a system change or if there is any mistake during pump selection, having the second largest impeller diameter in the system guarantees the possibility of changing the impeller to the next larger size without change the casing.

7.2 Centrifugal Fan

A centrifugal fan is a mechanical device for moving air or other gases, as shown in Fig. 7.5. It developed from simple paddle-wheel designs, in which the wheel was a disk carrying radial flat plates. Refinements have led to the three general types with backward-curved, radial-tipped, and forward-curved blades. All the fans illustrated have blades that are curved at their inlet edges to approximate shockless flow

Fig. 7.5 Centrifugal fan

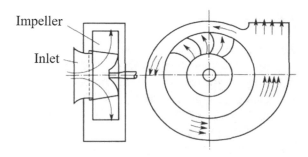

Impeller

Inlet

between the blade and the inlet flow direction. These three designs are typical of fans with sheet-metal blades, which are relatively simple to manufacture and thus relatively inexpensive.

The centrifugal fan increases the speed and volume of an air stream with the rotating impellers. Centrifugal fans use the kinetic energy of the impellers to increase the volume of the air stream, which in turn moves them against the resistance caused by ducts, dampers, and other components. Centrifugal fans displace air radially, changing the direction (typically by 90°) of the airflow.

7.2.1 Fan Performance Characteristics

As with the centrifugal pump, the centrifugal fan also has performance curves, the trend of which is similar to the centrifugal pump.

There also exists an operating point for centrifugal fan. The determination of the operating point is essential to make sure the fan functions normally.

7.2.2 Fan Selection and Installation

Selection and installation of a fan always requires compromise. To minimize energy consumption, it is desirable to operate a fan at its highest efficiency point. To reduce the fan size for a given capacity, it is tempting to operate at higher flow rate than that at maximum efficiency. In an actual installation, this tradeoff must be made considering such factors as available space, initial cost, and annual hours of operation. It is not wise to operate a fan at a flow rate below maximum efficiency. Such a fan would be larger than necessary and some designs, particularly those with forward-curved blades, can be unstable and noisy when operated in this region.

It is necessary to consider the duct system at both the inlet and the outlet of the fan to develop a satisfactory installation. Anything that disrupts the uniform flow at the fan inlet is likely to impair performance. Nonuniform flow at the inlet causes the wheel to operate unsymmetrically and may decrease capacity dramatically. Swirling flow also adversely affects fan performance. Swirl in the direction of rotation reduces the pressure developed; swirl in the opposite direction can increase the power required to drive the fan.

7.3   Problems

7.1   In a pumping station, the diameter of pump inlet is $d_1 = 250$ mm and $d_2 = 200$ mm (exit). The elevation height difference between the pressure gauge at the exit and the vacuum gauge at the inlet is $\Delta z = 0.3$ m. The vacuum gauge gives reading $p_v = 3.92 \times 10^4$ Pa, and the pressure gauge gives reading $p_M = 8.33 \times 10^5$ Pa. The flow rate is $Q = 60$ L/s. Find the head rise H of the pump.

7.2   A pump delivers saline water with flow rate $Q = 9000$ L/min and specific weight $\gamma = 11,760$ N/m^3. The diameter of pump exit is 250 mm, and at the inlet, the diameter is 300 mm. The centerlines of the inlet and the exit are on the same horizontal level, and the vacuum degree at the inlet is 150 mmHg . The pressure gauge, 1.2 m higher above the centerline of the exit, gives reading 1.4 atm. How much input power is required, in kW, if the efficiency is 0.84?

7.3   There is gasoline with specific weight $\gamma = 8330$ N/m^3 in the container, and at the bottom, there is an orifice connected with a steel pipe with diameter 50 mm. The gasoline is pumped out by a centrifugal pump and into an oil tank with same diameter steel pipe. The gasoline surface in the container and in the oil tank is 1.2 and 30 m higher than the axis of the pump, respectively. The overall length of the pipe is 38 m, and the flow rate is 4546 L/h. The dynamic viscosity of the gasoline is 0.8×10^{-3} Pa s. How much input power is required, in kW, if the efficiency is 0.80?

References

1. Karassick, I.J.: Pump Handbook, 3rd edn. McGraw-Hill, New York (2000)
2. Gülich, J.F.: Centrifugal Pumps, pp. 131–144. Springer, Berlin (2008)
3. Song, H.: Engineering Fluid Mechanics and Environmental Application. Metallurgical Industry Press, Beijing (2016)

Similitude and Dimensional Analysis

8

Abstract

Experiments are usually necessary to solve scientific and technologic problems because they can provide theoretical basis and criterion. The theoretical fundamentals for experimental design and evaluation are similitude and dimensional analysis. This chapter includes similitude, dimensional analysis, and its application. The learning goals are as follows. First, we should understand the concepts of mechanics similitude and similarity criterion. Then we have to master the applications of approximate similitude, the π theorem and dimensional analysis. Finally, three important approximate models should be grasped including Froude model, Reynolds model, and Euler model.

Keywords

Mechanics similitude · Froude number · Euler number · Approximate similitude · Pi theorem

8.1 Similitude

8.1.1 Basic Concepts of Mechanics Similitude

The experiments of engineering fluid mechanics mainly have two categories. One is engineering model experiment, aiming at hydraulic prediction in large machineries and upcoming constructed structures. The other is observation experiment, aiming at figuring out unknown flow characteristics. When investigating the inner mechanics and physical nature of fluid motion, all steps have to be based on scientific experiments, including proposing research method, developing fluid mechanics theory and solving engineering problems.

© Metallurgical Industry Press, Beijing and Springer Nature Singapore Pte Ltd. 2018
H. Song, *Engineering Fluid Mechanics*,
https://doi.org/10.1007/978-981-13-0173-5_8

To represent actual flow phenomena and predict flow properties utilizing physical model under actual conditions, it is necessary that prototype and model have mechanics similitude [1, 2]. Mechanics similitude indicates that all model variables at corresponding points are proportional to its prototype, and includes three aspects: geometric, kinematic and dynamic similitude.

(1) Geometric similitude

It means that the model and its prototype have identical shape, and differ only in scale.

Set the subscript p denote the prototype, and m denote the model. The length scale ratio λ_l (also named linear scale ratio) is defined as

$$\lambda_l = \frac{l_p}{l_m} \tag{8.1}$$

The area scale ratio λ_A and the volume scale ratio λ_V can be expressed as

$$\lambda_A = \frac{A_p}{A_m} = \frac{l_p^2}{l_m^2} = \lambda_l^2 \tag{8.2}$$

$$\lambda_V = \frac{V_p}{V_m} = \frac{l_p^3}{l_m^3} = \lambda_l^3 \tag{8.3}$$

The length scale ratio λ_l is the first basic scale ratio of mechanics similitude, which can be used to derive the area scale ratio λ_A and the volume scale ratio λ_V. The dimensions of length l, area A and volume V are L, L^2, and L^3, respectively.

(2) Kinematic similitude

It means that in addition to geometric similitude, velocities at all corresponding points are in the same ratio. The velocity scale ratio is

$$\lambda_v = \frac{v_p}{v_m} \tag{8.4}$$

It is the second basic scale ratio of mechanics similitude, and other kinematic scale ratios can be deduced with λ_l and λ_v according to the definition and dimension.

The time scale ratio is

$$\lambda_t = \frac{t_p}{t_m} = \frac{l_p/v_p}{l_m/v_m} = \frac{\lambda_l}{\lambda_v} \tag{8.5}$$

The acceleration scale ratio is

$$\lambda_a = \frac{a_p}{a_m} = \frac{v_p/t_p}{v_m/t_m} = \frac{\lambda_v}{\lambda_t} = \frac{\lambda_v^2}{\lambda_l} \tag{8.6}$$

The flow rate scale ratio is

$$\lambda_Q = \frac{Q_p}{Q_m} = \frac{l_p^3/t_p}{l_m^3/t_m} = \frac{\lambda_l^3}{\lambda_t} = \lambda_l^2 \lambda_v \tag{8.7}$$

(3)   Dynamic similitude

It means that the kinds of force on the model and its prototype are the same, and all corresponding forces are in the same ratio.

The density scale ratio is

$$\lambda_\rho = \frac{\rho_p}{\rho_m} \tag{8.8}$$

It is the third basic scale ratio of mechanics similitude, and others dynamic scale ratios can be deduced with λ_ρ, λ_l and λ_v according to the definition or dimension.

The mass scale ratio is

$$\lambda_m = \frac{m_p}{m_m} = \frac{\rho_p V_p}{\rho_m V_m} = \lambda_\rho \lambda_l^3 \tag{8.9}$$

The force scale ratio is

$$\lambda_F = \frac{F_p}{F_m} = \frac{m_p a_p}{m_m a_m} = \lambda_m \lambda_a = \lambda_\rho \lambda_l^2 \lambda_v^2 \tag{8.10}$$

The pressure (stress) scale ratio is

$$\lambda_p = \frac{F_p/A_p}{F_m/A_m} = \frac{\lambda_F}{\lambda_A} = \lambda_\rho \lambda_v^2 \tag{8.11}$$

Remarkably, the dimensionless scale ratio is

$$\lambda_C = 1 \tag{8.12}$$

It means that all dimensionless parameters equal to each other when the model and its prototype have mechanics similitude, which is able to measure actual flow parameters with its model such as velocity coefficient, flow rate coefficient and so on.

Meanwhile, the model and its prototype are usually in the same gravity field, so the scale ratio of gravity (gravitational acceleration exactly) generally equals 1, namely

$$\lambda_g = \frac{g_p}{g_m} = 1 \tag{8.13}$$

All these scale ratios of mechanics similitude are listed in Table 8.1. Basic scale ratios λ_l, λ_v and λ_ρ are independent to each other, so they can be used to deduce scale ratios of other parameters, then the relation between the model and its prototype can be determined.

8.1.2 Similarity Criterion

If the model and prototype flow have mechanics similitude, there must be lots of scale ratios. However, it is impossible and unnecessary to check scale ratios one by one to match mechanics similitude, and it can be easily done with similitude.

For incompressible fluid flow, the projection of differential motion equation in the x direction is

$$X - \frac{1}{\rho}\frac{\partial p}{\partial x} + v\nabla^2 u_x = \frac{du_x}{dt} \tag{8.14}$$

So all parameters in the model must be proportional to prototype flow if they have mechanics similitude, then the motion equation of prototype flow can be expressed as

$$\lambda_g X - \frac{\lambda_p}{\lambda_\rho \lambda_l}\frac{1}{\rho}\frac{\partial p}{\partial x} + \frac{\lambda_v \lambda_v}{\lambda_l^2} v\nabla^2 u_x = \frac{\lambda_v^2}{\lambda_l}\frac{du_x}{dt} \tag{8.15}$$

As we know that all terms in N-S equation have the dimension of acceleration LT^{-2}, so all scale ratios in the formula above are the acceleration scale ratio.

$$\lambda_g = \frac{\lambda_p}{\lambda_\rho \lambda_l} = \frac{\lambda_v \lambda_v}{\lambda_l^2} = \frac{\lambda_v^2}{\lambda_l} \tag{8.16}$$

The four terms in Eq. (8.16) all have definite physical meanings, which represent the ratio of gravity, pressure, viscous force, and inertial force between prototype flow and the model, respectively.

The fourth term is divided by the first three terms in Eq. (8.16) respectively, then we have three equations as follows:

Table 8.1 The scale ratios in different types of mechanics similitude

Types	Mechanics similitude	Gravity similarity Froude model	Viscous force similarity Reynolds model	Pressure similarity Euler model
Similarity criterion	$Fr_p = Fr_m$ $Re_p = Re_m$ $Eu_p = Eu_m$	$\dfrac{v_p^2}{g_p l_p} = \dfrac{v_m^2}{g_m l_m}$	$\dfrac{v_p l_p}{v_p} = \dfrac{v_m l_m}{v_m}$	$\dfrac{p_p}{\rho_p v_p^2} = \dfrac{p_m}{\rho_m v_m^2}$
Restriction between scale ratios	$\lambda_l \lambda_v \lambda_\rho$	$\lambda_v = \lambda_l^{\frac{1}{2}}$	$\lambda_v = \dfrac{\lambda_v}{\lambda_l}$	$\lambda_p = \lambda_\rho \lambda_v^2$
Linear scale ratio λ_l	Basic scale ratio	Basic scale ratio	Basic scale ratio	Same with 'Mechanics similitude' bar
Area scale ratio λ_A	λ_l^2	λ_l^2	λ_l^2	
Volume scale ratio λ_V	λ_l^3	λ_l^3	λ_l^3	
Velocity scale ratio λ_v	Basic scale ratio	$\lambda_l^{\frac{1}{2}}$	$\dfrac{\lambda_v}{\lambda_l}$	
Time scale ratio λ_t	$\dfrac{\lambda_l}{\lambda_v}$	$\lambda_l^{\frac{1}{2}}$	$\dfrac{\lambda_l^2}{\lambda_v}$	
Acceleration scale ratio λ_a	$\dfrac{\lambda_v^2}{\lambda_l}$	1	$\dfrac{\lambda_v^2}{\lambda_l^3}$	
Flow rate scale ratio λ_Q	$\lambda_l^2 \lambda_v$	$\lambda_l^{\frac{5}{2}}$	$\lambda_v \lambda_l$	
Kinematic viscosity scale ratio λ_v	$\lambda_l \lambda_v$	$\lambda_l^{\frac{3}{2}}$	Basic scale ratio	
Angular velocity scale ratio λ_ω	$\dfrac{\lambda_v}{\lambda_l}$	$\lambda_l^{-\frac{1}{2}}$	$\dfrac{\lambda_v}{\lambda_l^2}$	
Density scale ratio λ_ρ	Basic scale ratio	Basic scale ratio	Basic scale ratio	
Mass scale ratio λ_{lm}	$\lambda_\rho \lambda_l^3$	$\lambda_\rho \lambda_l^3$	$\lambda_\rho \lambda_l^3$	
Force scale ratio λ_F	$\lambda_\rho \lambda_l^2 \lambda_v^2$	$\lambda_\rho \lambda_l^3$	$\lambda_\rho \lambda_v^2$	
Moment scale ratio λ_M	$\lambda_\rho \lambda_l^3 \lambda_v^2$	$\lambda_\rho \lambda_l^4$	$\lambda_\rho \lambda_l \lambda_v^2$	
Energy scale ratio λ_E	$\lambda_\rho \lambda_l^3 \lambda_v^2$	$\lambda_\rho \lambda_l^4$	$\lambda_\rho \lambda_l \lambda_v^2$	
Pressure (stress) scale ratio λ_p	$\lambda_\rho \lambda_v^2$	$\lambda_\rho \lambda_l$	$\dfrac{\lambda_\rho \lambda_v^2}{\lambda_l^2}$	
Dynamic viscosity scale ratio λ_μ	$\lambda_\rho \lambda_l \lambda_v$	$\lambda_\rho \lambda_l^{\frac{3}{2}}$	$\lambda_\rho \lambda_v$	
Power scale ratio λ_P	$\lambda_\rho \lambda_l^2 \lambda_v^3$	$\lambda_\rho \lambda_l^{\frac{7}{2}}$	$\dfrac{\lambda_\rho \lambda_v^3}{\lambda_l}$	
Dimensionless scale ratio λ_C	1	1	1	

$$\frac{\lambda_v^2}{\lambda_g \lambda_l} = 1 \tag{8.17}$$

Or

$$\frac{v_{\mathrm{p}}^2}{g_{\mathrm{p}} l_{\mathrm{p}}} = \frac{v_{\mathrm{m}}^2}{g_{\mathrm{m}} l_{\mathrm{m}}} \tag{8.18}$$

$\frac{v^2}{gl} = Fr$ is called Froude number, which represents the ratio of inertial force to gravity.

$$\frac{\lambda_\rho \lambda_v^2}{\lambda_p} = 1 \text{ or } \frac{\lambda_p}{\lambda_\rho \lambda_v^2} = 1 \tag{8.19}$$

Or

$$\frac{p_{\mathrm{p}}}{\rho_{\mathrm{p}} v_{\mathrm{p}}^2} = \frac{p_{\mathrm{m}}}{\rho_{\mathrm{m}} v_{\mathrm{m}}^2} \tag{8.20}$$

$\frac{p}{\rho v^2} = Eu$ is called Euler number, which represents the ratio of pressure to inertial force.

$$\frac{\lambda_v \lambda_l}{\lambda_\nu} = 1 \tag{8.21}$$

Or

$$\frac{v_{\mathrm{p}} l_{\mathrm{p}}}{\nu_{\mathrm{p}}} = \frac{v_{\mathrm{m}} l_{\mathrm{m}}}{\nu_{\mathrm{m}}} \tag{8.22}$$

$\frac{vl}{\nu} = Re$ is called Reynolds number, which represents the ratio of inertial force to viscous force.

In a word, if two flows have mechanics similitude or complete similarity, Froude number, Euler number and Reynolds number must equal respectively for the prototype and its model, that is

$$\begin{cases} Fr_{\mathrm{p}} = Fr_{\mathrm{m}} \\ Eu_{\mathrm{p}} = Eu_{\mathrm{m}} \\ Re_{\mathrm{p}} = Re_{\mathrm{m}} \end{cases} \tag{8.23}$$

Equation (8.23) is the criterion of complete similarity for incompressible steady flow. It is obvious that similitude criterion utilization is more convenient than checking scale ratios one by one to match mechanics similitude. Similitude can be used not only to judge similarity but also to design models.

Mechanics similitude must be satisfied with three mutual restrictions on scale ratios as follows:

$$\begin{cases} \lambda_v^2 = \lambda_g \lambda_l \\ \lambda_p = \lambda_\rho \lambda_v^2 \\ \lambda_v = \lambda_l \lambda_v \end{cases} \qquad (8.24)$$

If the three basic scale ratios selected for model design can satisfy the three restrictions above, it means that the model and its prototype have complete mechanics similitude. However, there is challenge to satisfy Eq. (8.24) completely. For example, substituting the scale ratio of gravity $\lambda_g = 1$ into the first formula in Eq. (8.24), we have

$$\lambda_v = \lambda_l^{\frac{1}{2}} \qquad (8.25)$$

Substituting the equation above into the third formula in Eq. (8.24), we have

$$\lambda_v = \lambda_l^{\frac{3}{2}} \qquad (8.26)$$

In general, the model and its prototype use the same fluid (for example, aviation equipment tested in wind tunnel, hydraulic models tested with water, hydraulic component tested with working oil), so $\lambda_v = 1$, which leads to $\lambda_l = 1$. It does not make sense that the prototype and model flow have same length.

Due to certain restrictions among scale ratios, it is difficult to satisfy both Froude criterion and Reynolds criterion at the same time for model and its prototype flow. There is no contradiction between Euler criterion and the other two criterions above, so it is easy to reach the Euler criterion and one of two criterions, Froude criterion and Reynolds criterion, for model and its prototype flow. This approach which cannot have complete mechanics similitude is called incomplete similarity.

8.1.3 Incomplete Similarity/Approximate Similitude

Approximate similitude works in many cases. Froude number represents the ratio of inertial force to gravity, and Reynolds number represents the ratio of inertial force to viscous force. The three forces do not have the same influences on a certain problem, so it is feasible to reveal main contradiction after ignoring one of secondary factors and approximate similitude should focus on principal factors.

There are three approximate models in common use [3].

(1) Froude model

For fluid flow in hydraulic structures or open tunnels, gravity is the dominant force which plays key role in specific cases. For example, gravity dominates the flow from high position to low position due to elevation difference. Viscous forces have no or little impact on this kind of flow. The main similarity criterion of Froude model is

$$\frac{v_p^2}{g_p l_p} = \frac{v_m^2}{g_m l_m}$$

The model and prototype flow usually have the same gravitational acceleration, So

$$\frac{v_p^2}{l_p} = \frac{v_m^2}{l_m} \tag{8.27}$$

Or

$$\lambda_v = \lambda_l^{\frac{1}{2}} \tag{8.28}$$

This formula shows that the velocity scale ratio cannot be basic scale ratio in Froude model. Substituting Eq. (8.27) into relevant formulas from Eqs. (8.1) to (8.13), then the relationships between parameters and basic scale ratios λ_l and λ_ρ can be obtained (listed in Table 8.1).

Froude model is widely used in hydraulic engineering, and large hydraulic engineering must be tested with model experiments prior to construction.

(2) Reynolds model

Pipe flow needs to overcome pipe friction and is driven by pressure drop. Viscous force determines the pressure drop and flow properties, while gravity has little impact on flow and can be neglected. The main criterion of Reynolds model is

$$\frac{v_p l_p}{v_p} = \frac{v_m l_m}{v_m} \tag{8.29}$$

Or

$$\lambda_v = \frac{\lambda_v}{\lambda_l} \tag{8.30}$$

It demonstrates that the velocity scale ratio depends on the length scale ratio λ_l and the kinematic viscosity scale ratio λ_v. Substituting Eq. (8.30) into relevant formulas from Eqs. (8.1)–(8.13), the relationships between parameters and basic scale ratios λ_l, λ_v and λ_ρ can be obtained (listed in Table 8.1).

Reynolds model is widely used in model experiments on pipe flow, hydraulic technology, and hydraulic machinery.

(3) Euler model

As experiments show, when Reynolds number increases to a certain value, the influence of viscous force on flow relatively decreases. And it would have no longer significant influence on flow when Reynolds number continues to increase. Once viscous force and gravity have little impact on flow, we can utilize the Euler model.

The Euler criterion should satisfy $\lambda_p = \lambda_\rho \lambda_v^2$ and also select λ_l, λ_v and λ_ρ as the independent basic scale ratios. The scale ratios of other parameters are respectively the same with mechanics similitude while applying Euler criterion for designing model experiments.

Euler model is often used in pipe flows, wind tunnel with Reynolds number big enough.

Example 8.1 Figure 8.1 shows the water flow under tainter gate. The depth of water is $H = 4$ m.

(1) Try to determine the depth of water H' in the model while $\lambda_\rho = 1$ and $\lambda_l = 10$.
(2) The flow rate in the model is $Q_m = 155$ L/s, and the velocity of vena contracta is $v_m = 1.3$ m/s. The force and moment on the tainter gate are $F_m = 50$ N and $M_m = 70$ N m, respectively. Try to determine the flow rate, the velocity of vena contracta, the force and moment on the tainter gate in prototype flow.

Solution

The water flow under tainter gate is driven by gravity, so its model is designed according to Froude model, and the scale ratios can be obtained from Table 8.1.

(1) The water depth in the model

$$H' = \frac{H}{\lambda_l} = \frac{4}{10} = 0.4 \text{ m}$$

(2) For prototype

Flow rate

$$Q_p = \lambda_Q Q_m = \lambda_l^{\frac{5}{2}} Q_m = 10^{\frac{5}{2}} \times 0.155 = 49 \text{ m}^3/\text{s}$$

Fig. 8.1 Tainter gate

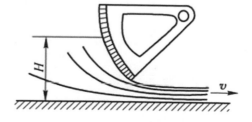

Velocity of vena contracta

$$v_p = \lambda_v v_m = \lambda_l^{\frac{1}{2}} v_m = \sqrt{10} \times 1.3 = 4.11 \text{ m/s}$$

Force on the tainter gate

$$F_p = \lambda_F F_m = \lambda_\rho \lambda_l^3 F_m = 1 \times 10^3 \times 50 = 5 \times 10^4 \text{ N}$$

Moment on the tainter gate

$$M_p = \lambda_M M_m = \lambda_\rho \lambda_l^4 M_m = 1 \times 10^4 \times 75 = 7.5 \times 10^5 \text{ N m}$$

Example 8.2 There is an oil pipe with diameter 15 cm and length 5 m. The flow rate is 0.2 m³/s. In the experiment, the oil is replaced by water. The pipes in the model and its prototype have same diameter. The oil viscosity is $v = 0.13 \text{ cm}^2/\text{s}$, and the water temperature is 10 °C. Find the flow rate in the model to make the two flows have similarity. If the pressure drop is 3 cm for a 5 m long pipe in the model, what is the pressure drop for a 100 m long pipe in its prototype?
Solution

(1) The main force of pipe flow is viscous force, so Reynolds numbers of two flows should equal to each other, namely, $Re_p = Re_m$, so $\lambda_Q = \lambda_v \lambda_l$.

Oil viscosity is $v_p = 0.13 \text{ cm}^2/\text{s}$, and water viscosity is $v_m = 0.0131 \text{ cm}^2/\text{s}$ at 10 °C (Table 1.2), so

$$\lambda_v = \frac{v_p}{v_m} = \frac{0.13}{0.0131} \approx 10.0$$

$$Q_m = \frac{Q_p}{\lambda_Q} = \frac{Q_p}{\lambda_v \lambda_l} = \frac{0.2}{10 \times 1} = 0.02 \text{ m}^3/\text{s}$$

(2)

$$\lambda_p = \frac{\lambda_\rho \lambda_v^2}{\lambda_l^2}$$

Since $\lambda_\gamma = \lambda_\rho \lambda_g$, the pressure of its prototype is

$$h_p = \left(\frac{\Delta p}{\gamma}\right)_p = h_m \lambda_p / \lambda_\gamma = h_m \lambda_v^2 / (\lambda_g \lambda_l^2)$$

While $\lambda_g = 1$ and $\lambda_l = 1$, if the pressure drop for a 5 m long pipe in the model is 0.03 m, the corresponding pressure drop in its prototype is

$$h_p = 0.03 \times (0.13/0.0131)^2/1 = 2.95 \text{ m}$$

For a 100 m long pipe in the prototype, the pressure drop is

$$2.95 \times 100/5 = 59 \text{ m}$$

8.2 Dimensional Analysis and Its Application

Fluid mechanics is more heavily involved with empirical work than is structural engineering and machine design, because the analytical tools presently available are not capable of yielding exact solutions to many of the problems in fluid mechanics. It is true that exact solutions are obtainable for all hydrostatic problems and for many laminar-flow problems. However, the most general equations solved on the largest computers yield only fair approximations for turbulent-flow problems-thus the need for experimental evaluation and verification. For analyzing model studies and for correlating the results of the experimental research, it is essential that the researchers employ dimensionless parameters.

8.2.1 Buckingham Pi Theorem

The Buckingham Pi theorem is a statement of the relation between a function expressed in terms of dimensional parameters and a related function expressed in terms of nondimensional parameters [4]. The Buckingham Pi theorem allows us to develop the important nondimensional parameters quickly and easily.

Assuming dimensional parameter N depending on dimensional parameters n_1, n_2, n_3 ... n_k, we have

$$N = f(n_1, n_2, n_3, \ldots, n_i, \ldots, n_k) \tag{8.31}$$

Select n_1, n_2, and n_3 as repeating parameters, which are a set of dimensional parameters that includes all the primary dimensions (e.g., L, M, T). Of course, the selection for the repeating parameters should meet two requirements. First, these parameters should be independent to each other; second, the dimensions of other parameters can be derived from these parameters.

For example, when solving problems about hydraulic head loss or flow resistance, it usually works out best to choose characteristic length l (dimension L in the MLT system), velocity v (dimensions LT^{-1}), and density ρ (dimensions ML^{-3}) as repeating parameters. These three parameters are independent and can respectively represent the feature of geometry, kinematics, and dynamics. So it is feasible to choose $n_1 = l$, $n_2 = v$, $n_3 = \rho$ as repeating parameters.

All the dimensional parameters in Eq. (8.31) can be expressed as the combination of repeating parameters and the dimensionless parameters as follows:

$$\begin{cases} N = \pi n_1^x n_2^y n_3^z \\ n_i = \pi_i n_1^{x_i} n_2^{y_i} n_3^{z_i} \end{cases} \tag{8.32}$$

The dimensionless parameters π and π_i are shown as

$$\begin{cases} \pi = \frac{N}{n_1^x n_2^y n_3^z} \\ \pi_i = \frac{n_i}{n_1^{x_i} n_2^{y_i} n_3^{z_i}} \end{cases} \tag{8.33}$$

The function between dimensionless parameters π and π_i also can correspondingly reflect the relationship between dimensional parameters N and n_i after transformation in terms of Eq. (8.33). Thus, Eq. (8.31) can be expressed as

$$\frac{N}{n_1^x n_2^y n_3^z} = f\left(\frac{n_1}{n_1^{x_1} n_2^{y_1} n_3^{z_1}}, \frac{n_2}{n_1^{x_2} n_2^{y_2} n_3^{z_2}}, \frac{n_3}{n_1^{x_3} n_2^{y_3} n_3^{z_3}}, \dots, \frac{n_i}{n_1^{x_i} n_2^{y_i} n_3^{z_i}}, \dots, \frac{n_k}{n_1^{x_k} n_2^{y_k} n_3^{z_k}} \right) \tag{8.34}$$

The first three items in the right side are repeating parameters, so we have

$$\begin{cases} x_1 = 1 & y_1 = z_1 = 0 \\ y_2 = 1 & x_2 = z_2 = 0 \\ z_3 = 1 & x_3 = y_3 = 0 \end{cases}$$

Then, Eq. (8.34) can be rewritten as

$$\pi = f(1, 1, 1, \pi_4, \pi_5, \dots, \pi_i, \dots, \pi_k)$$

Or

$$\pi = f(\pi_4, \pi_5, \dots, \pi_i, \dots, \pi_k) \tag{8.35}$$

Utilizing this new method, the previous function (8.31) with $k + 1$ dimensional parameters can be transformed into a function (8.35) with $k - 2$ dimensionless parameters. This is the Buckingham Pi theorem.

8.2.2 Applications of Dimensional Analysis

The five steps listed below outline a recommended procedure for determining the π parameters

Step 1: List all the dimensional parameters involved, that is $N, n_1, n_2, n_3 \ldots n_k$.
Step 2: Select a set of repeating parameters.
Step 3: Select a set of primary dimensions, such as L, M, T.
Step 4: List the dimensions of all parameters in terms of primary dimensions.
Step 5: Set up dimensional equations, combining the parameters selected in Step 2 with each of the other parameters in turn, to form dimensionless groups. (There will be $k - 2$ equations.) Solve the dimensional equations to obtain the $k - 2$ dimensionless groups.

The detailed procedure for determining the dimensionless π parameters is illustrated in Example 8.3.

Example 8.3 According to experiments, the pressure drop Δp caused by friction loss when flowing in the pipe is related to the following factors: pipe diameter d, average velocity v, density ρ, dynamic viscosity μ, pipe length l, and pipe surface roughness Δ. Try to determine the friction loss in the pipe.
Solution
It can be obtained according to the problem that

$$\Delta p = f(d, v, \rho, \mu, l, \Delta)$$

Select d, v and ρ as repeating parameters, because they meet two requirements for repeating parameters selection, then

$$\pi = \frac{\Delta p}{d^x v^y \rho^z}, \pi_4 = \frac{\mu}{d^{x_4} v^{y_4} \rho^{z_4}}, \pi_5 = \frac{l}{d^{x_5} v^{y_5} \rho^{z_5}}, \pi_6 = \frac{\Delta}{d^{x_6} v^{y_6} \rho^{z_6}}$$

Select L, M, T as primary dimensions and the dimensions of all parameters are listed as follows:

Parameters	d	v	ρ	Δp	μ	l	Δ
Dimensions	L	LT^{-1}	ML^{-3}	$ML^{-1}T^{-2}$	$ML^{-1}T^{-1}$	L	L

Set up dimensional equations as follows:
For Δp:

$$ML^{-1}T^{-2} = L^x (LT^{-1})^y (ML^{-3})^z = M^z L^{x+y-3z} T^{-y}$$

Thus $z = 1, y = 2, x = 0$
So

$$\pi = \frac{\Delta p}{v^2 \rho}$$

For μ:

$$ML^{-1}T^{-1} = L^{x_4}\left(LT^{-1}\right)^{y_4}\left(ML^{-3}\right)^{z_4} = M^{z_4}L^{x_4+y_4-3z_4}T^{-y_4}$$

Thus $z_4 = 1$, $y_4 = 1$, $x_4 = 1$
So

$$\pi_4 = \frac{\mu}{dv\rho}$$

Similarly

$$\pi_5 = \frac{l}{d}, \quad \pi_6 = \frac{\Delta}{d}$$

Substituting all dimensionless π parameters into Eq. (8.35), we have

$$\frac{\Delta p}{v^2\rho} = f\left(\frac{\mu}{dv\rho}, \frac{l}{d}, \frac{\Delta}{d}\right)$$

Since the friction loss in pipe flow is $h_f = \frac{\Delta p}{\rho g}$, and $Re = \frac{vd}{v} = \frac{vd\rho}{\mu}$, the above equation can be rewritten as

$$h_f = \frac{\Delta p}{\rho g} = \frac{v^2}{g}f\left(\frac{1}{Re}, \frac{l}{d}, \frac{\Delta}{d}\right)$$

According to experimental results, the friction loss is directly proportional to the pipe length l and inversely proportional to the pipe diameter d, so $\frac{l}{d}$ can be extracted from the above function. In addition, the contribution of Re to the function is equivalent to its reciprocal, so $\frac{1}{Re}$ can be replaced by Re. The above function can be transformed into

$$h_f = f\left(Re, \frac{\Delta}{d}\right)\frac{l}{d}\frac{v^2}{2g} = \lambda\frac{l}{d}\frac{v^2}{2g}$$

8.3 Problems

8.1 As shown in Fig. 8.2, the dimensions of a Venturi tube in the kerosene pipeline are $D = 300$ mm and $d = 150$ mm, and the flow rate is $Q = 100$ L/s. The kinematic viscosity of kerosene is $v = 4.5 \times 10^{-6}$ m^2/s, and its density is $\rho = 820$ kg/m^3. The dimension of a model is reduced to 1/3 of that in its

Fig. 8.2 Problem 8.1

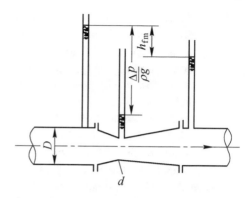

prototype and water is chosen for the test, of which the kinematic viscosity is $v_m = 1 \times 10^{-6}$ m²/s. Try to determine the flow rate in the model. If the head loss in the model is $h_{fm} = 0.2$ m and the pressure drop on the choke is $\Delta p_m = 10^5$ Pa, determining the corresponding head loss and pressure drop in the kerosene pipeline.

8.2 As shown in Fig. 8.3, the height of the car is $h = 2$ m, and its velocity is $v = 100$ km/h. The car is driven when the air temperature is 20 °C. The air temperature is 0 °C in the model, and the air velocity is $v' = 60$ m/s.

(1) Determine the car height h' in the model.
(2) If the resistance acting on the car in the model is $F' = 1500$ N, determining the corresponding resistance on the car in the prototype.

8.3 A torpedo is 5.8 m long, propelled in the sea (15 °C, $v = 1.5 \times 10^{-6}$ m²/s). The velocity of the torpedo is 74 km per hour. The torpedo model is 2.4 m long, tested in the water at 20 °C. Find the velocity of the model. If it is tested in the air in the standard state, what should the model velocity be?

8.4 Castor oil flows in the pipe with the velocity 5 m/s at 20 °C, and the diameter of the pipe is 75 mm. The density of castor oil is $\rho = 965$ kg/m³. The diameter of the pipe in the model is 50 mm, and air, as the model fluid, is in the standard state. To have dynamic similitude, what should the average velocity of air be?

Fig. 8.3 Problem 8.2

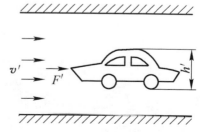

8.5 As shown in Fig. 8.4, investigating the flow over the top of the weir crest by utilizing length scale ratio $\lambda_l = 20$ in the experiment.

   (1) If the head over the weir crest in the prototype is $h = 3$ m, determining corresponding head in the model.

   (2) If the flow rate in the model is $Q_m = 0.19$ m^3/s, determining the flow rate in its prototype.

   (3) If the vacuum degree in weir crest in the model is $h_{vm} = 200$ mm, determining the corresponding vacuum degree in its prototype.

8.6 As shown in Fig. 8.5, utilizing water tower to simulate the pipe flow from kerosene tank. The viscosity of kerosene is $v = 4.5 \times 10^{-6}$ m^2/s, and the diameter of kerosene pipe is $d = 75$ mm. The water viscosity is $v_m = 1 \times 10^{-6}$ m^2/s. Try to determine:

   (1) diameter of water pipe; (2) scale ratio of liquid level height; (3) flow rate scale ratio.

8.7 Oil flows in the pipe with the diameter 20 cm. The oil viscosity is $v_p = 0.4$ cm^2/s, and the flow rate is 121 L/s. If the diameter of the pipe in the model is 5 cm, assuming that the model fluid is: (1) water at 20 °C ($v_m = 1.003 \times 10^{-6}$ m^2/s) (2) air ($v_m = 0.17$ cm^2/s), then determining the flow rate in the model respectively. Assuming the main force is viscous force.

8.8 The ratio of ventilation tunnel to its model is 30:1. In the model, the water dynamic viscosity is 50 times the air viscosity, and its density is 800 times the

Fig. 8.4 Problem 8.5

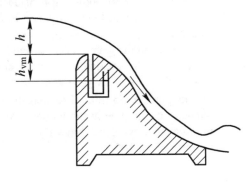

Fig. 8.5 Problem 8.6

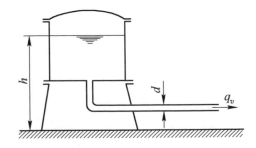

air density. The model and the prototype have dynamic similitude. If the
pressure drop in the model is 22.8×10^4 N/m^2, what should the pressure drop
be in the prototype expressed in mm H$_2$O?

8.9   As shown in Fig. 8.6. The flow rate per length in the rectangular weir is
$\frac{Q}{B} = kH^x g^y$, where k is a constant, H is the head in weir crest, g is gravitational
acceleration. Try to determine the values of x and y using dimensional
analysis.

8.10 As shown in Fig. 8.7. The flow rate through the orifice is related to diameter d
of the orifice, fluid pressure p and density ρ. Try to determine the function of
flow rate using dimensional analysis.

8.11 When liquid flows in the pipes which have geometric similitude, the pressure
drop can be expressed as $p = \frac{\rho l v^2}{d} \phi \left(\frac{vd\rho}{\mu} \right)$. Try to prove it. ($d$ is pipe diameter, l
is pipe length, ρ is fluid density, μ is dynamic viscosity, v is the fluid velocity,
ϕ represents a function.)

8.12 The resistance acting on the board is R, which is submerged and moving in a
fluid. It is known that the resistance is related to factors as follows: density ρ
and viscosity μ of the fluid, velocity v, length l and width b of the board. Try to
obtain the expression for the resistance.

8.13 The input power of the fan is related to the impeller diameter D, rotation
angular velocity ω and fluid viscosity μ. Try to determine the relationship
between the input power and other parameters using dimensional analysis.

8.14 The resistance F acting on the small ball is related to its velocity v in the fluid,
the ball diameter D, fluid density ρ and dynamic viscosity μ. Try to express the
resistance in a function with relevant parameters using dimensional analysis.

Fig. 8.6 Problem 8.9

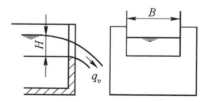

Fig. 8.7 Problem 8.10

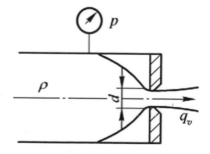

References

1. Howarth, L.: Modern Developments in Fluid Dynamics: high Speed Flow, vol. 1. Claredon Press, Oxford (1953)
2. Heller, V.: Scale effects in physical hydraulic engineering models. J. Hydraul. Res. **49**(3), 293–306 (2011)
3. Song, H.: Engineering Fluid Mechanics and Environmental Application. Metallurgical Industry Press, Beijing (2016)
4. Fox, R.W., McDonald, A.T., Pritchard, P.J.: Introduction to fluid mechanics, vol. 5, 8th edn. Wiley, New York (2011)

CO$_2$ Storage in Saline Aquifer with Vertical Heterogeneity

9

Abstract

CCS (CO$_2$ Capture and Storage) is a classical application of fluid flow in porous media. In recent years, CCS technology is considered an effective way to reduce CO$_2$ emissions. Saline aquifer is given special affection for CCS because of its huge amount of storage. In this chapter, we will first discuss mathematics model for CO$_2$ storage in terms of fluid flow theory, then introduce analytical solution to obtain percolating resistance and sweep efficiency. Finally, we evaluate and analyze storage effect of CO$_2$.

Keywords

Carbon capture and storage (CCS) · Vertical heterogeneity · Analytical solutions · Percolating resistance · Sweep efficiency

9.1 Description of CO$_2$ Storage

The process of CO$_2$ (Carbon dioxide) storage in nature has been happening for millions of years. But the artificial way to inject CO$_2$ into the formation has been implemented since 1970s. CO$_2$ storage is one of the core contents of CCS (CO$_2$ Capture and Storage) technology, and also the most challenging chain in the whole process of CCS. At present, the geological structures suitable for CO$_2$ storage include deep saline aquifer, exhausted oil and gas field, and basalt aquifer [1].

Massive emissions of CO$_2$ into the atmosphere are the most direct reason that causes global warming and climate change, so more and more countries are starting to focus on carbon abatement technologies. In recent years, the method, CCS is considered the most effective way to reduce greenhouse gas emissions [2, 3]. Saline aquifer is given special affection because of its huge amount of storage. Therefore,

© Metallurgical Industry Press, Beijing and Springer Nature Singapore Pte Ltd. 2018
H. Song, *Engineering Fluid Mechanics*,
https://doi.org/10.1007/978-981-13-0173-5_9

Table 9.1 List of global CCS projects

Project name, Nation	CO$_2$ processing capacity Mt/a (year)	Geological structure	Sequestration depth
Sleipner, Norway	1 (1996)	Marine saline aquifer	1000
In Salah, Algeria	1 (2004)	Continental saline aquifer	1850
Frio Project, U.S.A	0.65 (2004)	Continental saline aquifer	1540
Snohvit, Norway	0.75 (2008)	Marine saline aquifer	2600
Gorgon, Australia	3.3 (2009)	Marine saline aquifer	2500
Oedos, China	>0.1 (2010)	Continental saline aquifer	3000
West Pearl Queen, U.S.A	0.036 (2002)	Exhausted oil field	1370
Otway Basin, Australia	1 (2005)	Exhausted oil field	2056
Total Lacq, France	0.075 (2006)	Exhausted oil field	4500

deep saline aquifer is the best choice for the storage of CO$_2$ [4, 5]. At present, the representative projects of CCS, which are running or building in the world, is shown in Table 9.1.

Generally, saline aquifer injected with CO$_2$ is located deeper than 800 m underground where CO$_2$ is in supercritical state (pressure of 7.382 MPa, temperature of 31.048 °C is the supercritical state of CO$_2$) [6]. This condition can improve the efficiency of the CO$_2$ storage. Therefore, the process of injecting and storing CO$_2$ in saline aquifers is the process of the two-phase displacement of supercritical CO$_2$ and brine. Supercritical CO$_2$-brine two-phase displacement process is very complex, involving two-phase flow in porous media, two-phase interface moves, and other physical processes.

9.2   Mathematics Model for CO$_2$ Storage in Saline Aquifer

9.2.1   Continuity Equations of Two-Phase Flow

During the injection of CO$_2$, brine flow out from porous media due to the pressure. Continuity equation in the percolation process is as below. Area 1 gradually expands as time progresses.

(1) The continuity equations of the supercritical CO$_2$ and brine in the i-th layer region 1 are

$$-\nabla \cdot \left[\rho_{gi} \cdot \bar{V}_{gi}\right] = \frac{\partial}{\partial t}\left(\phi_i \rho_{gi} S_{gi}\right) \tag{9.1a}$$

$$-\nabla \cdot \left[\rho_{wi} \cdot \bar{V}_{wi}\right] = \frac{\partial}{\partial t}\left(\phi_i \rho_{wi} S_{wi}\right), \tag{9.1b}$$

where ρ_{gi}, ρ_{wi} are the density of CO$_2$ and brine respectively, kg/m^3; V_{gi}, V_{wi} are the percolation velocity vector of CO$_2$ and brine respectively in layer i; S_{gi}, S_{wi} are the saturation of CO$_2$ and brine in i-layer region; ϕ_i is the porosity of i-layer saline aquifer.

The continuity equation of i-layer region 2 is shown in Eq. (9.1b).

(2) According to the hypothesis, without considering the variation of compressibility and density of supercritical CO$_2$ and brine, we can get the continuity equation of mixing zone and brine zone in i-th layer through Eq. (9.1a, 9.1b):

$$-\nabla \cdot V_{gi} = \phi_i \frac{\partial S_{gi}}{\partial t} \tag{9.2a}$$

$$-\nabla \cdot V_{wi} = \phi_i \frac{\partial S_{wi}}{\partial t} \tag{9.2b}$$

In polar coordinates, Eq. (9.2a, 9.2b) can be written as

$$\frac{1}{r_i} \cdot \frac{\partial}{\partial r}\left(r_i V_{gi}\right) = -\phi_i \frac{\partial S_{gi}}{\partial t} \tag{9.3a}$$

$$\frac{1}{r_i} \cdot \frac{\partial}{\partial r}\left(r_i V_{wi}\right) = -\phi_i \frac{\partial S_{wi}}{\partial t} \tag{9.3b}$$

9.2.2   Momentum Equations of Two-Phase Flow

According to the assumptions, CO$_2$ and brine follow Darcy's law in the displacement process without considering the effect of gravity and capillary forces. Therefore, the momentum equation of the mixing zone can be expressed as

$$v_{gi} = -\frac{k_i \cdot k_{rgi}}{\mu_{gi}} \nabla p_i \tag{9.4a}$$

$$v_{wi} = -\frac{k_i \cdot k_{rwi}}{\mu_{wi}} \nabla p_i \tag{9.4b}$$

Momentum equation of the brine region can be expressed as

$$v_{wi} = -\frac{k_i}{\mu_{wi}} \nabla p_i, \tag{9.5}$$

where k_{rgi}, k_{rwi} are the relative permeability of CO_2 and brine respectively in mixing zone of layer i; k_i is the permeability of saline aquifer in i-th layer; μ_{gi}, μ_{wi} are the viscosity of CO_2 and brine respectively in layer i; p_i represents the pressure of the layer i, Pa.

9.2.3 State Equations

(1) Saturation of saline aquifer

Seen from the assumption, the mixture of CO_2 and brine is completed instantaneously in porous media during the process of continual injection of CO_2, and the sum of saturations equals to one, namely

$$S_{gi} + S_{wi} = 1 \tag{9.6}$$

(2) Fractional equation of supercritical CO_2

Fractional equation is expressed with the ratio of CO_2 volumetric flow rate to the total flow, expressed by f_g. Through the assumptions of neglecting the impact of the capillary forces and gravitational forces, the fractional equation of the i-th layer can be represented by the equation

$$f_{gi} = \frac{q_{gi}}{q_{gi} + q_{wi}} = \frac{v_{gi}}{v_{gi} + v_{wi}} = \frac{1}{1 + \frac{\mu_{gi}}{\mu_{wi}} \cdot \frac{k_{rwi}}{k_{rgi}}} \tag{9.7}$$

where f_{gi} is the CO_2 content of the i-th layer.

From Eq. (9.7) we know that, without considering the influence of gravity and capillary forces, the fractional equation mainly depends on the ratio of the viscosity and permeability between CO_2 and brine. For a particular saline aquifer, in the process of CO_2 injection, the value of μ_{gi}, μ_{wi} are almost unchanged. Therefore, the variation of f_g is mainly affected by k_{rwi}/k_{rgi}, and the relative permeability is a function of saturation. Consequently, f_g is also a function of saturation.

(3) Relative permeability

Permeability saturation curve is important to predict CO_2 storage efficiency. Permeability saturation curves measured in different hierarchical saline aquifers are different from each other. Based on experience, we can use the equation to get the permeability saturation curve and related data. In saline aquifer, the permeability saturation curves of CO_2 and brine can be expressed as

$$k_{rgi} = k_{rgi\,max} \left(\frac{S_{gi} - S_{rgi}}{1 - S_{rgi} - S_{rwi}} \right)^{\alpha_i} \tag{9.8a}$$

$$k_{rwi} = k_{rwi\,max} \left(\frac{1 - S_{gi} - S_{rwi}}{1 - S_{rgi} - S_{rwi}} \right)^{\beta_i} \tag{9.8b}$$

where, S_{rgi}, S_{rwi} are the residual saturation of CO_2 and brine in i-th layer, respectively; $k_{rgi\,max}$, $k_{rwi\,max}$ are the maximum relative permeability of CO_2 and brine in the i-th layer; α_i, β_i are constant relating to the rock pore structure in the i-th layer of saline aquifer.

We can obtain the values of α, β by linear regression, obtain average value of the relative permeability of different testing rock samples α, β, S_{rg}, S_{rw} by using the arithmetic average method, and substitute into Eq. (9.8a, 9.8b) where you can find permeability saturation curves of supercritical CO_2 and brine in different saline aquifers.

9.2.4 Governing Equation

According to Eqs. (9.2a, 9.2b), (9.6), and (9.7), the displacement process could be described by

$$\frac{dr_i}{dt} = \frac{q_i(t)}{2\pi h_i \phi_i r_i} \frac{df_{gi}}{dS_{gi}} \tag{9.9}$$

where h_i refers to the average thickness of saline aquifer of the i-th layer; $q_i(t)$ is the flow rate into the i-th layer with the variation of time.

Through variable separation and integration of Eq. (9.9), the saturation distribution of CO_2 in saline aquifers can be obtained, namely

$$\int_{r_0}^{r} r_i dr = \frac{f'_{gi}}{2\pi h_i \phi_i} \int_{0}^{t} q_i(t) dt$$

$$r_i^2 - r_0^2 = \frac{f'_{gi}}{\pi h_i \phi_i} W_i(t) \tag{9.10}$$

where r_i refers to front position radius in the i-th layer saline aquifers at the time t: $W_i(t) = \int_0^t q_i(t)dt$ refers to the total amount of CO_2 injected into the i-th layer at the time t; r_0 is the radius of the injection well; f'_{gi} refers to $f'_{gi} = df_{gi}/dS_{gi}$.

As the single flow $q_i(t)$ changes with time, the front radius for each time can be determined after iterations and the calculation of the time period.

9.3 Analytical Solution

According to the mathematical model, combined with geological conditions of saline aquifers and the injection conditions we can obtain the expression of motion law and storage efficiency of the supercritical CO_2 in saline aquifers.

9.3.1 The Front Saturation and the Average Saturation

(1) The front saturation

By means of experiments and analysis, due to the influence of the presence of gravity and capillary forces, the front saturation is gradually and slowly changing. In engineering, it is handled as mutation values to meet the accuracy requirements.

From the above analysis we know that the amount of CO_2 injected into a saline aquifer at time t is the increments of saturation, namely

$$\int_0^t q_i(t)dt = \int_{r_0}^r 2\pi h_i \phi_i r_i \left[S_{gi}(r_i, t) - S_{rgi} \right] dr_i \tag{9.11}$$

Substitute $dr_i = \frac{W_i(t)}{2\pi h_i \phi_i r_i} f''_{gi} \cdot dS_{gi}$ into Eq. (9.11),

$$\int_{S_{g0i}}^{S_{gfi}} \left[S_{gi}(r_i, t) - S_{rgi} \right] f''_{gi} dS_{gi} = 1 \tag{9.12}$$

where S_{g0i} is the saturation of CO_2 when $r = r_0$; S_{gfi} is the saturation of CO_2 at the front position.

Saturation of CO_2 that injected into the borehole wall is $S_{g0} = 1$. When the saturation of CO_2 is 1, namely $f_g(1) = 1$, $f'_g(1) = 0$, from Eq. (9.12) we obtain

$$f'_{gi}(S_{gfi}) = \frac{f_{gi}(S_{gfi})}{S_{gfi} - S_{rgi}} \tag{9.13}$$

Equation (9.13) contains the implicit function of S_{gf}. We can use the mapping method to obtain f'_{gi}, and then substitute it into Eq. (9.10) to obtain the curve of the growth of the front radius with time.

(2) **The average saturation**

By the assumptions, in the same saline aquifer, when CO_2 is injected into saline aquifer, its average saturation keeps unchanged in the displacement process. After obtaining the front saturation of the mixing zone, we can further determine the average saturation $\bar{S}_{gi}$, namely

$$\bar{S}_{gi} - S_{rgi} = \frac{q_i(t) \cdot t}{\pi h_i (r_{fi} - r_0)^2 \phi_i} \tag{9.14}$$

Substituting Eq. (9.10) into (9.14),
$\bar{S}_{gi} - S_{rgi} = \frac{r_{fi} + r_0}{r_{fi} - r_0} \cdot \frac{1}{f'_{gi}(S_{gi})}$, As the radius of saline aquifer R is much greater than the radius of the well radius r_0, we can ignore the radius of the well and get

$$\bar{S}_{gi} - S_{rgi} = \frac{1}{f'_{gi}(S_{gi})} \tag{9.15}$$

According to the significance of Eq. (9.15), the average saturation of the CO_2 injecting process of the i-th layer can also be obtained by mapping method.

9.3.2 Two-Phase Displacement Interface Movement

By Eq. (9.10) we obtain the expression of movement of the front radius with the injection of the supercritical CO_2 as follows:

$$r_i(t) = \sqrt{\frac{f'_{gi}}{\pi h_i \phi_i} W_i(t) - r_0^2} \tag{9.16}$$

$W_i(t)$ of the i-th layer saline aquifer changes with time, and other parameters can be obtained from known conditions, so the front radius function $r_i(t)$ is a function of time. Only by redistributing the flow in the calculation cycle can we get the position of the front radius at this moment.

9.3.3 Injection Pressure

In the assumptions, CO_2 is injected into saline aquifers with the constant flow rate. In the i-th layer of saline aquifers, the flow of the mixing zone consists of CO_2 and brine, namely

$$Q_i(t) = q_i(t) + q_{wi} = -\frac{k_i k_{rgi}}{\mu_{gi}} 2\pi r_i h_i \frac{dp}{dr} - \frac{k_i k_{rwi}}{\mu_{wi}} 2\pi r_i h_i \frac{dp}{dr}$$
$$= -2k_i \pi r_i h_i \left(\lambda_{gi} + \lambda_{wi}\right) \frac{dp}{dr},$$

(9.17)

where, Q_i is the flow rate of the i-th layer of saline aquifers at time t; k_i is absolute permeability of the i-th layer of saline aquifers; λ_{gi}, λ_{wi} are the conductivity of CO$_2$ and brine respectively, which is defined as $\lambda = k_r/\mu$.

Through variable separation of Eq. (9.17), we get

$$\int_P^{P_r} dp = -\frac{Q_i(t)}{2k_i \pi h_i \left(\lambda_{gi} + \lambda_{wi}\right)} \int_{r_0}^{r_i} \frac{1}{r_i} dr$$

$$P(t) - P_r = \frac{Q_i(t) \cdot \ln\frac{r_i}{r_0}}{2k_i \pi h_i \left(\lambda_{gi} + \lambda_{wi}\right)},$$

(9.18)

where $P(t)$ is the injection pressure at the injection well and it is a function of time; P_r is the pressure at the position of the front; Similarly, the pressure in the brine zone can be expressed as

$$Q_i(t) = -\frac{k_i}{\mu_{wi}} 2\pi r_i h_i \frac{dp}{dr},$$

After integration: $\int_{P_r}^{P_h} dp = \frac{Q_i(t)\mu_{wi}}{2\pi h_i k_i} \int_r^R \frac{1}{r_i} dr,$

$$P_r - P_h = \frac{Q_i(t)\mu_{wi}}{2\pi h_i k_i} \ln\frac{R}{r_i}$$

(9.19)

In the equation, P_h is the hydrostatic pressure of saline aquifers; R is the radius of the storage area.

Simultaneously solving Eqs. (9.18), (9.19) we can get the expression of the injection pressure as follows:

$$P(t) - P_h = \frac{Q_i(t)}{2\pi h_i k_i} \left(\mu_{wi} \cdot \ln\frac{R}{r_i} + \frac{\ln\frac{r_i}{r_0}}{\left(\lambda_{gi} + \lambda_{wi}\right)}\right)$$

(9.20)

9.3.4   Percolating Resistance

Regard the injection system as the circuit system, namely the so-called electronic–hydraulic analogy. With flow rate corresponding to current flow and differential pressure corresponding to voltage, we can obtain from Eq. (9.20)

$$Q_i(t) = \frac{P(t) - P_h}{\frac{\ln\frac{r_i}{r_0}}{2\pi h_i k_i \cdot (\lambda_{gi} + \lambda_{wi})} + \frac{\mu_{wi} \cdot \ln\frac{R}{r_i}}{2\pi h_i k_i}} \tag{9.21}$$

From the above equation we know that percolating resistance is composed of two parts, the mixing zone resistance (F_{1i}) and brine zone resistance (F_{2i}). Defining percolating resistance as F, so the mixing zone, the brine zone, and the total percolating resistance can be expressed as

$$F_{1i} = \frac{1}{2\pi h_i k_i \cdot (\lambda_{gi} + \lambda_{wi})} \cdot \ln\frac{r_i}{r_0} \tag{9.22a}$$

$$F_{2i} = \frac{\mu_{wi}}{2\pi h_i k_i} \cdot \ln\frac{R}{r_i} \tag{9.22b}$$

$$F_i = F_{1i} + F_{2i} = \frac{\ln\frac{r_i}{r_0}}{2\pi h_i k_i \cdot (\lambda_{gi} + \lambda_{wi})} + \frac{\mu_{wi} \cdot \ln\frac{R}{r_i}}{2\pi h_i k_i} \tag{9.23}$$

In the equation, F_{1i}, F_{2i} are the percolating resistance of the i-th layer in zone 1 and zone 2 respectively, Pa; F_i is the total resistance of the i-th layer.

Equation (9.23) is the expression of the flow resistance of supercritical CO_2. In the model, fluidity, permeability, thickness, injection radius are given, the only parameter that changes is the front radius, which changes with time and is related to percolating resistance. So the change of front radius with time can be analyzed by iteration.

9.3.5 CO₂ Flow Rate of Each Layer

Known from percolating resistance, the resistance of each layer corresponds to the circuit resistance in circuitry. Therefore the total resistance of the whole saline aquifer corresponds to the resistance of the layers in parallel. Namely

$$F_t = \prod_{i=1}^{n} F_i \Bigg/ \sum_{i=1}^{n} \frac{\prod_{i=1}^{n} F_i}{F_i} \tag{9.24}$$

F_t is the total resistance of the entire saline aquifers.

Resistance changes with time. In the beginning of each computation cycle, the flow rate of each layer is reassigned as follows, so that the flow rate is a function of time.

$$q_i(t) = qF_t/F_i \tag{9.25}$$

9.3.6 Heterogeneity Coefficient

To describe the differences between the various layers, we need to define hetero-geneity coefficient between layers to describe the heterogeneity between layers. The heterogeneity coefficient refers to the ratio between mean square error and average permeability of the permeability of each layer in statistical intervals [7, 8]. The greater the heterogeneity coefficient of permeability is, the stronger the hetero-geneity between layers. It is defined as follows:

$$\bar{k} = \frac{\sum_{i=1}^{n} h_i k_i}{\sum_{i=1}^{n} h_i} \tag{9.26}$$

$$K_v = \frac{\sqrt{\sum_{i=1}^{n} \left(k_i - \bar{k}\right)^2 \big/ n}}{\bar{k}} \tag{9.27}$$

In the equation: $\bar{k}$ is the average permeability of saline aquifers; h_i refers to the effective thickness of the i-th layer. K_v refers to the heterogeneity coefficient between layers.

Generally, the formation is uniform when $K_v \leq 0.5$, relatively uniform when $0.5 \leq K_v \leq 0.7$, and heterogeneous when $K_v \geq 0.7$.

9.3.7 Sweep Efficiency

In the process of the storage of CO_2 in saline aquifers, storage efficiency is an important indicator to evaluate the geological storage. In this study, defining sweep efficiency is used to evaluate the storage efficiency of CCS. Sweep efficiency is defined as follows:

$$\eta = \frac{\sum_{i=1}^{n} \pi r_i^2 h_i}{\pi \cdot r_{\max}^2 H} = \frac{\sum_{i=1}^{n} r_i^2 h_i}{r_{\max}^2 H} \tag{9.28}$$

In the equation, η is the sweep efficiency; $r_{\max}$ is the maximum front radius of the n-th layer saline aquifer; H represents the total thickness of the saline aquifer.

Sweep efficiency is mainly related to the front radius during the CO_2 injection process. Under the same injection conditions, the greater the permeability of the formation is, the larger the front radius is. $r_{\max}$ refers to the front radius of the layer with the largest permeability.

9.3.8 Calculation Processes

According to the established supercritical CO_2-brine two-phase displacement model, using the software Matlab, calculation process is shown in Fig. 9.1.

Figure 9.1 is a calculation process flow chart. According to Eq. (9.8a, 9.8b), the relative permeability curves and the relative permeability of supercritical CO_2 and brine in each layer can be derived. The initial allocation of flow in each layer is determined by the absolute permeability.

Taking one day as a computation cycle, the position of the two-phase fluid displacement interface within this period can be obtained by Eq. (9.16). When the front radius is fixed, the percolating resistance and sweep efficiency within this period can be determined by Eqs. (9.23) and (9.28). Based on the percolating resistance of each layer, we can redistribute flow rate by Eq. (9.25), and then enter the next cycle until the end of the cycle. At last the data is output, and drawn as a diagram.

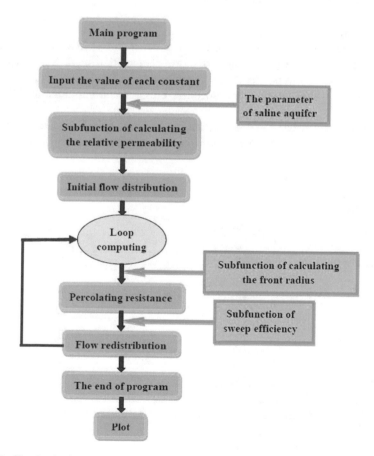

Fig. 9.1 Sketch of calculating process

9.4   Evaluation and Analysis of Storage Effect

Resistance of the process of CO_2 injection, sweep efficiency, and the impact of various factors are analyzed respectively. As constant flow rate is used in supercritical CO_2 injection process, the variation law of injection pressure could be described by multiplying percolating resistance by flow rate. Therefore injection pressure will not be discussed in the following discussion.

Sweep efficiency is used as the parameter to quantify storage efficiency. The higher the sweep efficiency is, the higher the storage efficiency of saline aquifer is. Studies of effect of influential factors, including saline aquifer radial extent, injection rate, heterogeneity coefficient, porosity, and other controllable injection parameters as well as geological parameters, provide important theoretical basis for the design optimization of CO_2 storage.

9.4.1   Scheme and Parameter Selection

9.4.1.1   Parameter Selection

A larger number of layers n will result in a better approximation to the actual situation of the saline aquifer, and yet more computation load. Suited for engineering needs, three layers, that is, $n = 3$, were assumed in the calculation in order to study the CO_2 injection process, injection pressure and variation law of sweep efficiency.

The parameters needed in the calculation are shown in Table 9.2. Schematic model is shown in Fig. 9.2. Mt/a means that injecting a million tons into the saline aquifer annually.

Table 9.2 Parameters

Parameters	Symbol	Unit	Values
Saline aquifer porosity	ϕ	%	35
Saline aquifer thickness	h_1	m	30
	h_2		35
	h_3		40
Brine viscosity	μ_w	Pa·s	1.33×10^{-4}
Brine density	ρ_w	kg/m^3	1100
CO_2 fluid viscosity	μ_g	Pa·s	5.8×10^{-5}
CO_2 fluid density	ρ_g	kg/m^3	760
Injection rate	q	Mt/a	0.5
Residual brine saturation	S_{rw}	%	30
Residual CO_2 saturation	S_{rg}	%	10
Temperature in aquifer	T	°C	35
Radial extent	R	km	20

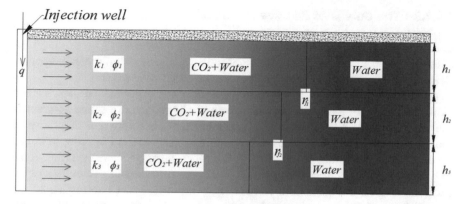

Fig. 9.2 The sketch of each brine layer

Permeability has effect on the interface movement of the two-phase fluid displacement. As shown in Fig. 9.2, the greater the permeability, the larger the moving distances. The difference of the two-phase displacement interface positions will affect the storage efficiency.

The parameters in Table 9.2 are based on experimental data.

9.4.1.2 Scheme

Such influential parameters were considered as permeability of saline aquifer, radial extent, porosity, injection rate, and injection pressure, among which the first three are the physical characteristics of saline aquifer, and the latter two belong to the injection conditions. Heterogeneity coefficient defined in Eq. (9.27) is the index to represent the heterogeneity of saline aquifer, related to the thickness and permeability of each layer. To simplify the calculations, the thicknesses of three saline aquifers are assumed to be constant.

The schemes listed in Table 9.3 include different values of heterogeneity coefficient and injection rate. In subsequent calculations, the values of other parameters are based on Table 9.2.

Table 9.3 Different schemes

	Case number	Formation permeability (m²)			Heterogeneity coefficient	Injection rate (Mt/a)
		Layer 1	Layer 2	Layer 3		
Base case	Case 1	6.0×10^{-14}	4.5×10^{-14}	3.0×10^{-14}	0.2803	0.5
Formation permeability	Case 2	7.7×10^{-14}	4.3×10^{-14}	1.9×10^{-14}	0.5498	0.5
	Case 3	8.8×10^{-14}	4.4×10^{-14}	1.0×10^{-14}	0.7370	0.5
	Case 4	11×10^{-14}	2.0×10^{-14}	1.4×10^{-14}	1.0165	0.5
Injection condition	Case 5	6.0×10^{-14}	4.5×10^{-14}	3×10^{-14}	0.2803	0.1
	Case 6	6.0×10^{-14}	4.5×10^{-14}	3×10^{-14}	0.2803	0.3
	Case 7	6.0×10^{-14}	4.5×10^{-14}	3×10^{-14}	0.2803	1

9.4.1.3 The Curves of Relative Permeability

The relative permeability of supercritical CO_2 and brine is related to saline water saturation in the layer and residual saturation of the injected CO_2, varying with different CO_2 saturation. The expression of the relative permeability of supercritical CO_2 and saline water in saline aquifer is shown as Eq. (9.8a, 9.8b).

According to geological data of a saline aquifer and experimental data, we can draw relative permeability curves and curves of CO_2 content in saline aquifers as shown in Fig. 9.3. As shown in Fig. 9.3, relative permeability and derivative of supercritical CO_2 content with respect to saturation are required in order to obtain front radius. The meaning of Eq. (9.13) is that when we make the tangent line of the CO_2 saturation curve through the point $S_g = 0.1$, the corresponding saturation of the cut point is the front saturation. According to Eq. (9.15), the intersection of the tangent line and the line $y = 1$ is the average saturation, from which the relative permeability of brine and supercritical CO_2 during CO_2 injection and the average saturation of supercritical CO_2 can be obtained.

9.4.2 Influential Factors of Percolating Resistance

Percolating resistance describes the magnitude of the resistance of the percolation received in the formation. In the process of the injection of CO_2 into the formation, the less the percolating resistance is, the faster the injection efficiency will be in the

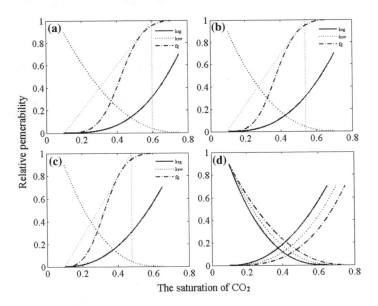

Fig. 9.3 The curves of relative permeability and the saturation of CO_2: **a** the first layer; **b** the second layer; **c** the third layer; **d** three-layer comparison chart of permeability curve

same injection pressure. Therefore, in this study, the analysis of the influencing factors of percolating resistance has great effect on the improvement of injection efficiency and storage efficiency.

9.4.2.1   Variation Law of Percolating Resistance

According to the scheme Base case listed in Table 9.3, the heterogeneity coefficient is 0.2803, the injection is 0.5 Mt/a, the variation of percolating resistance of the three layers with time is shown in Fig. 9.4.

As shown in the Fig. 9.4, percolating resistance of the three layers is increasing with time. It presents an evident upward trend in the initial stages and tends to be stabilized over time. In the Base Case, the absolute permeability of each layer in the saline aquifer from top to bottom are: 60 mD; 45 mD; 30 mD, and the heterogeneity coefficient 0.2803, which indicates the formation is homogeneous. However, the percolating resistance of each layer is quite different. As shown in the Fig. 9.4, the greater the absolute permeability is, the smaller the percolating resistance will be. The percolating resistance of the uppermost layer increases from the initial 130–135 gradually. The percolating resistance of the intermediate layer increases from the initial 153–160 gradually. And the percolating resistance of the lowest layer increases from the initial high 204–220 and tends to stability.

Based on the above analysis, with the increase of the absolute permeability in the stratum, the percolating resistance is decreasing. Besides, with the linear decrease of the permeability, the percolating resistance is increasing linearly.

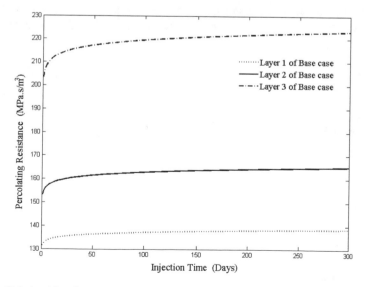

Fig. 9.4 Relationship of each brine layer's percolating resistance with the base case

9.4.2.2 Influencing Factors of Percolating Resistance

According to Eq. (9.20) we know that the pressure gradient is the product of the total percolating resistance and flow rate. When the injection rate is given, the greater the pressure is, the more energy is consumed and the more cost will be. Variation of percolating resistance has important effect on improving storage efficiency, economic efficiency and cost reduction. According to Eq. (9.23) of percolating resistance, this section will study the effect on percolating resistance of various factors.

(1) **Injection rate of supercritical CO$_2$**

Injection rate will directly affect the efficiency of injecting CO$_2$ into the saline aquifer. The higher the injection rate is, the more conducive to the storage of CO$_2$ and the more reduction of greenhouse effect. However, the higher the injection rate is, the greater the resistance is. The curve of the total resistance of supercritical CO$_2$-brine two-phase fluid flow in the percolating process is shown in Fig. 9.5.

Injection rates were among 0.1, 0.3, 0.5, 1 Mt/a. Seen from the Fig. 9.5, the total resistance increases with different growth rates under different conditions, and reaches a plateau afterwards. With lower injection rate, the flow resistance is smaller and it is more conducive to inject CO$_2$; increasing injection rate will lead to higher flow resistance and is more detrimental to the injection of CO$_2$.

(2) **Heterogeneity coefficient of the saline aquifer**

Heterogeneity coefficient indicates the heterogeneity of saline aquifers. The larger the heterogeneity coefficient is, the more heterogeneous the saline aquifer is. And

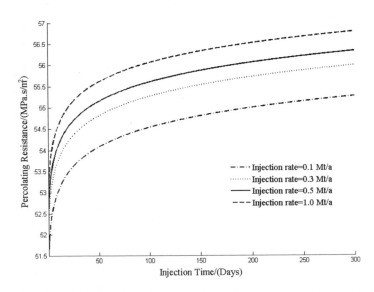

Fig. 9.5 The relationship of total percolating resistance changed at different injection rate

the smaller the heterogeneity coefficient is, the more homogenous the saline aquifer is and more conducive to CO_2 storage. Figure 9.6 shows the curve of the variation of percolating resistance with time with different heterogeneity coefficients. Figure 9.7 shows the curve of the variation of total percolating resistance with different heterogeneity coefficients.

Heterogeneity coefficient of Fig. 9.6 are 0.2803, 0.5498, 0.7370, 1.0165 indicating the increase of heterogeneity. Seen from the figure, the smaller the heterogeneity coefficient is, the greater the percolating resistance will be. With the increase of heterogeneity coefficient, the total percolating resistance is decreasing.

Figure 9.7 is the curve of the variation of percolating resistance with different heterogeneity coefficients. Seen from the figure, it can be divided into three regions. In the first region, the percolating resistance and sweep efficiency are great. Although we can take full advantage of the saline aquifer to store CO_2, the great percolating resistance is not conducive to CO_2 injection as it acquires high energy consumption. In the third region, the percolating resistance is low, but the sweep efficiency is also low, so the saline aquifer cannot be fully used and the storage efficiency is low. The second region is the moderate. The saline aquifer with heterogeneity coefficients increasing from 0.18 to 0.62 has benefit for CO_2 storage.

(3) Radial extent

Radial extent is an important factor when selecting storage site. The impact of different radial extents on percolating resistance is shown in Fig. 9.8 based on other parameters of Base case.

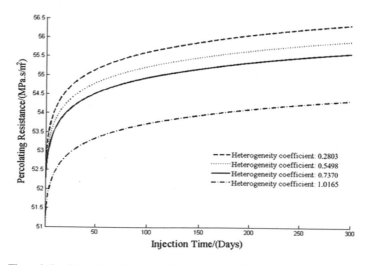

Fig. 9.6 The relationship of total percolating resistance changed with the time at different heterogeneity coefficient

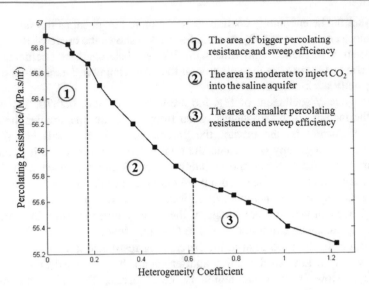

Fig. 9.7 The relationship of total percolating resistance changed with the heterogeneity coefficient

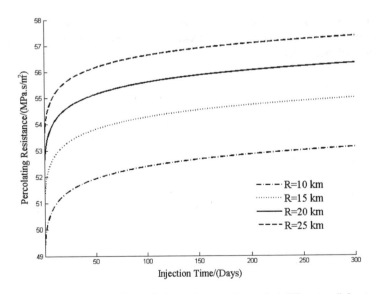

Fig. 9.8 The relationship of total percolating resistance changed at different radial extent

As seen from the figure, with the increase of the radius, percolating resistance is also gradually increasing. With the linear increase of radial extent, the total percolating resistance is not increasing linearly but very slowly. And the rate of increase leveled off over time.

(4) **Porosity of the saline aquifer**

Porosity of saline aquifers is an important parameter to evaluate the capacity of CO_2 storage in saline aquifer. The larger the porosity is, the greater the storage capacity will be. The impact of porosity on percolating resistance is shown in Fig. 9.9.

Figure 9.9 is a curve of the variation of percolating resistance with different porosity. The larger the porosity is, the smaller the percolating resistance will be and more conducive to CO_2 storage.

Figures 9.5–9.9 shows the curves of the variation over time of percolating resistance with different injection efficiency, heterogeneity coefficient, radial extent, and porosity. As seen from the figures, the total percolating resistance is increasing over time. At the initial period, it increases quickly, and it tends to increase slowly.

9.4.3 Influential Factors of Sweep Efficiency

The definition of sweep efficiency is shown in Eq. (9.28) in Sect. 9.3.7, and is an important indicator to evaluate CO_2 storage efficiency in the saline aquifer. Due to the formation heterogeneity, the advancing rates of CO_2 in different layers are different. This study analyzes the variation law of front radius and sweep efficiency under different values of heterogeneity coefficient, injection rate, radial extent, porosity to provide theoretical guidance for the CCS program.

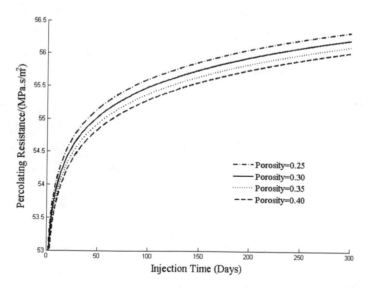

Fig. 9.9 The relationship of percolating resistance changed at different porosity

9.4.3.1 Sweep Efficiency and Front Radius

Heterogeneity coefficient defined by Eqs. (9.26) and (9.27) is a parameter to describe the heterogeneity between layers. According to Table 9.3, the heterogeneity coefficients are 0.2803, 0.5498, 0.7370, 1.0165, respectively.

Figure 9.10 shows the curves of variation of sweep efficiency with heterogeneity coefficient over time and the sketch map of front radius in saline aquifer. The four diagrams (a), (b), (c), (d) represent the variation of sweep efficiency over time and the corresponding sketch map of front radius when the injection efficiency in saline aquifer are 0.1, 0.3, 0.5, 1 Mt/a from (a) to (d) respectively. As can be seen from the figure, with the injection of supercritical CO$_2$, the sweep efficiency is decreasing. At the initial stage it decreases quickly and then level off.

Sweep efficiency decreases with the increase of the injection rate. We can see from the sketch map of front radius that with the increase of injection efficiency, the gaps between the layers become larger and larger and thus influence the CO$_2$ storage efficiency. Sweep efficiency also decreases with the increase of heterogeneity coefficient.

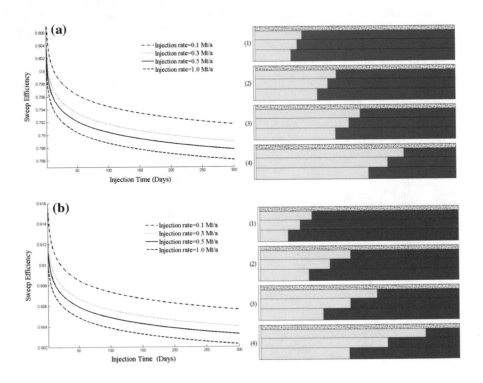

Fig. 9.10 The variation of sweep efficiency and front radius with heterogeneity coefficient in saline aquifer: **a** heterogeneity coefficient is 0.2803; **b** heterogeneity coefficient is 0.5498; **c** heterogeneity coefficient is 0.7370; **d** heterogeneity coefficient is 1.0165

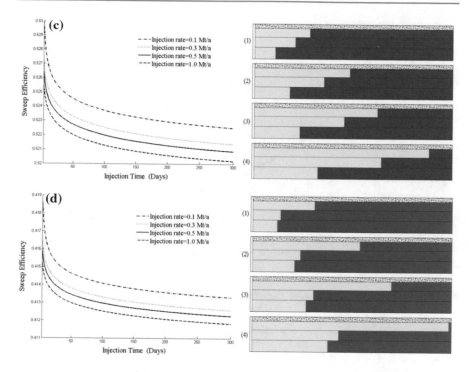

Fig. 9.10 (continued)

As can be seen from sketch map of front radius, when the heterogeneity coefficient is 1.0165, front radius of the first layer is much larger than that the other two layers, which causes a great waste for the saline aquifer that stores supercritical CO_2.

9.4.3.2   The Influence of Injection Rate

Figure 9.11 is the curve graph of the sweep efficiency with the variation of injection rate when the heterogeneity coefficient is 0.2803. As can be seen from Fig. 9.11, the sweep efficiency decreases from the previous 79.25–78.6% with the increases of injection rate, and the decreasing amplitude is only 0.65%. We can conclude that the injection rate has little impact on sweep efficiency.

9.4.3.3   The Influence of Heterogeneity Coefficient

From the above analysis we know that the larger the heterogeneity coefficient is, the smaller the percolating resistance will be and more conducive to the injection of CO_2. The influence of heterogeneity coefficient on CO_2 storage efficiency is shown as below.

Figure 9.12 shows the curve of the sweep efficiency with the variation of heterogeneity coefficient. As can be seen from the figure, with the increase of the

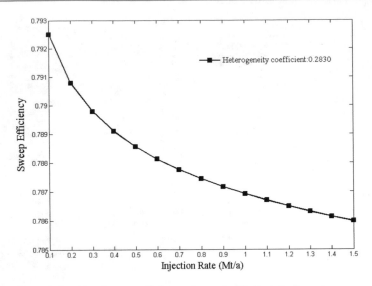

Fig. 9.11 The relationship of sweep efficiency changed with the injection rate

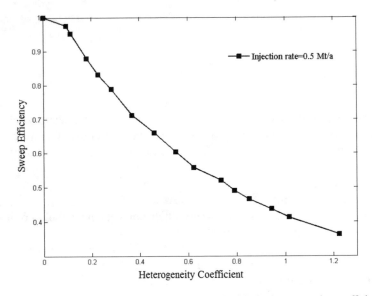

Fig. 9.12 The relationship of sweep efficiency changed with the heterogeneity coefficient

heterogeneity coefficient, the sweep efficiency decreases. The heterogeneity coefficient increases from the 0.1 to 1.2 and the sweep efficiency decreases from 95 to 35%. So we can see that heterogeneity coefficient is the factor that influence sweep efficiency most. Therefore, in the process of the evaluation of CO$_2$ storage in saline aquifer, the smaller the heterogeneity coefficient is, the higher the CO$_2$ storage efficiency will be and the better the economic efficiency.

Figures 9.10–9.12 represents the law of the variation of sweep efficiency with different heterogeneity coefficient and different injection rate over time. As can be seen from the figures, the sweep efficiency is declining over time. In the initial stage, the drop is large, and at last it tends to level off. Based on the above analysis we can see that the injection rate has little effect on sweep efficiency and heterogeneity coefficient, which shows the heterogeneity of the stratum itself, is the most important factor affecting sweep efficiency. The larger the heterogeneity coefficient is, the lower the sweep efficiency will be. The smaller the heterogeneity coefficient is, the higher the sweep efficiency will be. What's more, the influence of the injection rate on sweep rate also cannot be overlooked.

9.4.3.4 The Influence of Different Radial Extent

In the selection of the site for CO_2 storage, the larger the radial extent is, the more the storage of CO_2. But the radial extent of saline aquifer will influence the percolating resistance and indirectly influence the sweep efficiency. The analysis of the influence of different radial extent on sweep rate has great significance in the evaluation of storage efficiency of saline aquifer. In the calculation of the supercritical CO_2-Brine two-phase displacement model, the radial extent takes the value 20 km. We can choose different radial extents under this condition. The curve of the variation of sweep efficiency over time is shown in Fig. 9.13.

Seen from the figure, the larger the radial extent is, the higher the sweep efficiency will be. It indicates that in the selection of storage site, the larger the radius is, the more favorable of CO_2 storage. Radial extent can be taken as an important indicator for the evaluation of CCS.

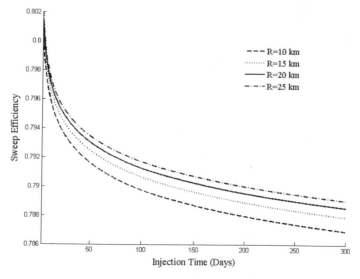

Fig. 9.13 The relationship of sweep efficiency changed with the time at different radial extent

9.4.3.5   The Influence of Porosity

In the evaluation of storage efficiency, porosity is a factor that cannot be over-looked. The larger the porosity, the more CO_2 is stored in the unit volume of saline aquifer and more favorable for CO_2 storage. With different porosity, the curve of the variation of sweep efficiency over time is shown in Fig. 9.14.

Seen from the figure, the larger the porosity is, the higher the sweep efficiency will be, and the more conducive to storage of CO_2. In the evaluation of CO_2 storage in saline aquifers this parameter is an important indicator for site selection.

Through the above analysis, the effect of influential factors on the total perco-lating resistance and sweep efficiency are clarified. With the increase of the heterogeneity coefficient, the percolating resistance and sweep efficiency are decreasing. The smaller the percolating resistance is, the more conducive to CO_2 storage. However, the decrease of storage efficiency will result in decrease of amount of CO_2 storage. Figure 9.7 shows that heterogeneity coefficient in the intermediate zone, ensure the CO_2 storage efficiency as well as quick decrease of the percolating resistance. An optimum selection of heterogeneity coefficient can benefit both the economic efficiency and injection efficiency of CO_2 storage.

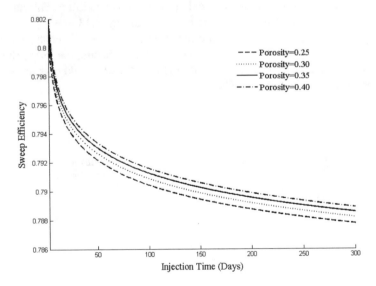

Fig. 9.14 The relationship of sweep efficiency changed with the time at different porosity

References

1. Song, H.: Engineering Fluid Mechanics and Environmental Application. Metallurgical Industry Press, Beijing (2016)
2. Javaheri, M., Jessen, K.: Integration of counter-current relative permeability in the simulation of CO_2 injection into saline aquifers. Int. J. Greenhouse Gas Control 5(5), 1272–1283 (2011)
3. Song, H., Huang, G., Li, T., Zhang, Y., Lou, Y.: Analytical model of CO_2 storage efficiency in saline aquifer with vertical heterogeneity. J. Nat. Gas Sci. Eng. 18, 77–89 (2014)
4. Alonso, J., Navarro, V., Calvo, B.: Flow path development in different CO_2 storage reservoir scenarios: a critical state approach. Eng. Geol. 127, 54–64 (2012)
5. Celia, M.A., Nordbotten, J.M., Dobossy, M., Bachu, S.: Field-scale application of a semi-analytical model for estimation of CO_2 and brine leakage along old wells. Int. J. Greenhouse Gas Control 5(2), 257–269 (2011)
6. Frailey, S.M.: Methods for estimating CO_2 storage in saline reservoirs. Energy Procedia 1(1), 2769–2776 (2009)
7. Deng, H., Stauffer, P.H., Dai, Z., Jiao, Z., Surdam, R.C.: Simulation of industrial-scale CO_2 storage: multi-scale heterogeneity and its impacts on storage capacity, injectivity and leakage. Int. J. Greenhouse Gas Control 10, 397–418 (2012)
8. Geiger, S., Schmid, K.S., Zaretskiy, Y.: Mathematical analysis and numerical simulation of multi-phase multi-component flow in heterogeneous porous media. Curr. Opin. Colloid Interface Sci. 17(3), 147–155 (2012)

Appendix

Chapter 1

1.1 $v = 1.52 \times 10^{-5}\,\text{m}^2/\text{s}$

1.2 $\mu = 1.87 \times 10^{-5}\,\text{Pa}\cdot\text{s}, v = 1.69 \times 10^{-5}\,\text{m}^2/\text{s}$

1.3 $\tau = 145.8\,\text{Pa}$

1.4 (1) $F_1 = 6\,\text{N}$ (2) $F_2 = 43.2\,\text{N}$

1.5 $F = 3.73\,\text{N}$

1.6 $39.6\,\text{N}\cdot\text{m}$

1.7 $V = 151.34\,\text{m}^3$

1.8 $p_2 = 172.2\,\text{kPa}$

1.9 $\Delta V = 0.2\,\text{m}^3$

Chapter 2

2.1 $p = 248.3\,\text{kPa}$

2.2 $p_0 = 104.24\,\text{kPa}$

2.3 Absolute pressure: $11.8\,\text{m}\,\text{H}_2\text{O}$, $0.868\,\text{m}\,\text{Hg}$; Relative pressure: $1.5\,\text{m}\,\text{H}_2\text{O}$, $110.34\,\text{mm}\,\text{Hg}$

2.4 $p' = -2940\,\text{Pa}; p_v = 2940\,\text{Pa}$

2.5 $p_M = 177.74\,\text{kPa}; p'_M = 76.44\,\text{kPa}$

2.6 $h_2 = 43.5\,\text{cm}$

2.7 $\Delta p = 27.32\,\text{kPa}$

2.8 $p' = 823.2\,\text{Pa}$

2.9 $p'_A = 264.8\,\text{kPa}$

2.10 (1) $p_B - p_A = 0.415\,\text{m}\,\text{H}_2\text{O}$ (2) $h_1 = 0.55\,\text{m}, h_2 = -0.2\,\text{m}, h_3 = 0.75\,\text{m}, z = 0.4\,\text{m}$

(3) $p_B - p_A = 0.4\,\text{m}\,\text{H}_2\text{O}$ (4) $h_1 = 0.95\,\text{m}, h_2 = -1\,\text{m}, h_3 = 0.35\,\text{m}$

2.11 $p_1 = 1.187 \times 10^3\,\text{kPa}$

2.12 $P = 27.15\,\text{kN}$

© Metallurgical Industry Press, Beijing and Springer Nature Singapore Pte Ltd. 2018
H. Song, *Engineering Fluid Mechanics*,
https://doi.org/10.1007/978-981-13-0173-5

2.13 $P = 25.58\,\text{kN}, h_D = 1.55\,\text{m}$
2.14 $P = 12.05\,\text{kN}, h_D = 1.60\,\text{m}$
2.15 $P = 1.02 \times 10^8\,\text{N}, Y = 27.9\,\text{m}$

Chapter 3

3.1 $y = 0, z = \frac{1}{x^2}$
3.2 $y - y_0 = \frac{x^2 - x_0^2}{1 + At_0}$
3.3 $a = 104i + 154j$
3.4 $a = 4i + 6j, a = 7.21\,\text{m/s}^2$
3.5 $Q = 61.36\,\text{m}^3/\text{s}, v = 78.13\,\text{m/s}$
3.6 $v = 0.5\,u_{max}$
3.7 $u_z = -xz + \frac{z^2}{2}$
3.8 $v_B = 4.5\,\text{m/s}, v_D = 10.88\,\text{m/s}$
3.9 $v_1 = 18.05\,\text{m/s}, v_2 = 22.25\,\text{m/s}$
3.10 $v_1 = 8.04\,\text{m/s}, v_8 = 6.98\,\text{m/s}$
3.11 $v_1 = 9.6\,\text{m/s}, Q_1 = 2.4\,\text{m}^3/\text{s}; v_2 = 6.4\,\text{m/s}, Q_2 = 1.6\,\text{m}^3/\text{s}; v_3 = 3.2\,\text{m/s}, Q_3 = 0.8\,\text{m}^3/\text{s}$
3.12 $Q_M = 0.518\,\text{kg/s}$
3.13 $u_0 = 49.50\,\text{m/s}$
3.14 $v_2 = 5.6\,\text{m/s}, d_2 = 5\,\text{cm}$
3.15 $v_A = 6\,\text{m/s}, h_1 = 1.73\,\text{m H}_2\text{O}$, from A to B
3.16 (1) $h_1 = -0.239\,\text{moil}$ (2) From 2-2 to 1-1 (3) $\Delta p = 3.74 \times 10^4\,\text{Pa}$
3.17 $Q = 0.091\,\text{m}^3/\text{s}$
3.18 $Q = 1.935\,\text{m}^3/\text{s}$
3.19 $v_3 = 10.84\,\text{m/s}, Q = 3.41\,\text{L/s}, p_2 = 22.83\,\text{kPa}$
3.20 $h_v = 47.3\,\text{mm Hg}$
3.21 $Q = 0.0512\,\text{m}^3/\text{s}$
3.22 $p = 53.66\,\text{kPa}$
3.23 $v = 5.72\,\text{m/s}, p_c = -68.99\,\text{Pa}$ (Relative pressure)
3.24 $v = 7.92\,\text{m/s}, H = 0.8\,\text{m}, d' = 80.6\,\text{mm}$
3.25 $H = 127.4\,\text{m}, N = 433.51\,\text{kW}$
3.26 $F = 9.46\,\text{kN}, \theta = 18.22°$
3.27 $F_x = 0.243\,\text{kN}, F_y = 0.026\,\text{kN}$
3.28 $y = 5.61\,\text{m}$
3.29 $F = 126\,\text{N}$
3.30 $\alpha = 30°, R = 456.5\,\text{N}$
3.31 (1) $F_1 = 920\,\text{N}$ (2) $F_2 = 230\,\text{N}$
3.32 $R_x = \rho A_1(v_1 - v)^2(\cos\theta - 1), R_y = \rho A_1(v_1 - v)^2\sin\theta, R = \sqrt{R_x^2 + R_y^2}$

Chapter 4

4.1 $Re = 1914 < 2000$ Laminar flow, $Re = 4787$ Turbulent flow, $Q = 5.12 \times 10^{-5}\,\text{m}^3/\text{s}$

4.2 $Re = 9289 > 300$ Turbulent flow; $v < 0.16\,\text{cm/s}$

4.3 $\mu = 0.135\,\text{Pa} \cdot \text{s}$

4.4 $\lambda = 4.65 \times 10^{-7}\,Re$

4.5 Turbulent flow in smooth pipes (1) $\lambda = 0.0236$ (2) $\delta = 1.88\,\text{mm}$ (3) $\tau_0 = 5.06\,\text{Pa}$

4.6 (1) $p = \gamma l \frac{\lambda h/d - 1}{\lambda h/d + 1}$ (2) $h + \frac{d}{\lambda}$ (3) $v = \sqrt{\frac{2g(h+l)}{1 + \lambda l/d}}$ (4) $h = \frac{d}{\lambda}$ (5) $p_A = 0$, $p_1 = p_2 = p_3 = p_4 = 0$

4.7 (1) Region IV: Turbulent transient region (2) $\lambda = 0.025$

4.8 $h_f = 40.82\,\text{moil}$

4.9 $h_f = 16.54\,\text{moil}$

4.10 $N = 19.53\,\text{kW}$

4.11 According to Blasius formula $\lambda = 0.015$, from Moody chart $\lambda = 0.019$

4.12 $\frac{D}{d} = \sqrt{2}$, $\Delta H_{\max} = \frac{v_d^2}{4g}$

4.13 (1) $h_f = 0.284\,\text{m}$ (2) $\frac{p_2 - p_1}{\gamma} = 0.455\,\text{m}$ (3) $\frac{p_2 - p_1}{\gamma} = 0.74\,\text{m}$

4.14 $h_r = 0.268\,\text{m}\,H_2O$

4.15 $\zeta = 0.65$

4.16 $v = 27.43\,\text{m/s}$

4.17 $\zeta = 6.275$

4.18 $h = 3.59\,\text{m}$

4.19 $Q = 53.5\,\text{L/s}$

4.20 $H = 26.6\,\text{m}$, $Q = 4.9\,\text{L/s}$

4.21 $H = 3.84\,\text{m}$

4.22 (1) $Q = 0.045\,\text{m}^3/\text{s}$ (2) $H_1 = 17.76\,\text{m}$

Chapter 5

5.1 $Q = 0.757\,\text{m}^3/\text{s}$

5.2 $Q = 13.57\,\text{L/s}$, $d = 135\,\text{mm}$

5.3 $h_{f1} = 0.824\,\text{m}\,H_2O$, $h_{f2} = 1.76\,\text{m}\,H_2O$, $h_{f3} = 9.42\,\text{m}\,H_2O$

5.4 $d_2 = 150\,\text{mm}$

5.5 $h_f = 7.39\,\text{m}\,H_2O$

5.6 $Q = 11\,\text{L/s}$

5.7 $d = 51.3\,\text{cm}$

5.8 (1) $Q_1 = 23.75\,\text{L/s}$, $Q_2 = 5.7\,\text{L/s}$ (2) $Q = 19.67\,L/s$

5.9 $H = 0.92\,\text{m}$

5.10 $Q = 0.132\,\text{m}^3/\text{s}$

5.11 $y = \frac{H}{2}$, $x_{\max} = C_v H$

5.12   (1) $\varphi = 0.92$, $\varepsilon = 0.65$ (2) $Q = 49.4\,\text{L/s}$
5.13 $v = 9.4\,\text{m/s}$, $Q = 6.66\,\text{L/s}$
5.14 $t = 144\,\text{s}$
5.15 $t = 18.5\,\text{h}$

Chapter 6

6.1   (C) 2.4 m/d
6.2 $v = 0.0637\,\text{cm/s}$, $u = 0.318\,\text{cm/s}$
6.3 $k = 0.6\,\text{m/s}$
6.4 $Q = 1.25\,\text{L/s}$

Chapter 7

7.1 $H = 89.41\,\text{m}$
7.2 $N = 31.1\,\text{kW}$
7.3 $N = 387.5\,\text{kW}$

Chapter 8

8.1 $Q_2 = 7.4\,\text{L/s}$, $h_{f1} = 0.45\,\text{m}$, $\Delta p_1 = 1.845\,\text{bar}$
8.2   (1) $h' = 0.873$ m (2) $P_1 = 1830\,\text{N}$
8.3 $v_{mw} = 120\,\text{km/h}$, $v_{ma} = 1585\,\text{km/h}$
8.4 $v_m = 0.103\,\text{m/s}$
8.5   (1) $h = 0.15\,\text{m}$ (2) $Q = 339.88\,\text{m}^3/\text{s}$ (3) $h = 4\,\text{m}\,H_2O$
8.6   (1) $d = 27.5\,\text{mm}$ (2) $\lambda_h = 2.73$ (3) $\lambda_Q = 12.27$
8.7   (1) $Q_m = 76\,\text{mL/s}$ (2) $Q_m = 12.87\,\text{L/s}$
8.8 $\Delta p = 8.25\,\text{mm}\,H_2O$
8.9 $x = \frac{3}{2}$, $y = \frac{1}{2}$
8.10 $Q = kd^2\sqrt{\dfrac{p}{\rho}}$

8.12 $Q = kd^2\left(\dfrac{\Delta p}{\rho}\right)^{\frac{1}{2}}$
8.13 $N = D^5\rho\omega^3 f\left(\dfrac{Q}{D^3\omega}\right)$
8.14 $F = \rho v^2 D^2 f(\text{Re})$

Printed in the United States
By Bookmasters